Lecture Notes in Physics

Edited by H. Araki, Kyoto, J. Ehlers, München, K. Hepp, Zürich
R. Kippenhahn, München, H. A. Weidenmüller, Heidelberg,
J. Wess, Karlsruhe and J. Zittartz, Köln
Managing Editor: W. Beiglböck

303

P. Breitenlohner D. Maison
K. Sibold (Eds.)

Renormalization of Quantum Field Theories with Non-linear Field Transformations

Proceedings of a Workshop, Held at Ringberg Castle
Tegernsee, FRG, February 16–20, 1987

Springer-Verlag
Berlin Heidelberg GmbH

Editors

Peter Breitenlohner
Dieter Maison
Klaus Sibold
Max-Planck-Institut für Physik und Astrophysik
Werner-Heisenberg-Institut für Physik
Föhringer Ring 6, D-8000 München 40, FRG

ISBN 978-3-662-13667-6 ISBN 978-3-540-39178-4 (eBook)
DOI 10.1007/978-3-540-39178-4

Originally published by Springer-Verlag Berlin Heidelberg New York in 1988
Softcover reprint of the hardcover 1st edition 1988

2158/3140-543210

Preface

Renormalization theory has had its major successes in the consistent formulation of local gauge theories. Whereas the computational methods for quantum electrodynamics (QED) are by now well established on the textbook level and are supported by the well-known high precision tests of low-energy QED, the corresponding calculus for the standard model has not yet developed to the same degree and is still a topic reserved for a relatively small group of specialists. Apart from the greater complexity of the standard model, this is due to the non-linear structure of the invariance transformations characterizing the geometrical content of such theories. The renormalization scheme independent solution of the quantization problem for the non-linear BRS (Becchi-Rouet-Stora) transformations led to startling connections with very abstract branches of mathematics (algebraic topology, cohomology theory of Lie algebras) and provided a fruitful stimulus for a refreshing dialogue between physicists and mathematicians. It was soon recognized that similar problems arise in the quantization of other field theories possessing a geometrical structure, as general relativity, supersymmetric Yang-Mills theories, supergravity theories and non-linear σ-models.

Frequently the elegant and abstract approach based on the methods of algebraic topology is not a very practicable one for actual calculations. Hence there has developed in parallel a 'subculture' enhancing the art of doing practical calculations using specific (supposedly invariant) renormalization schemes. Although this avoids the sophisticated reasoning of the algebraic approach, it has its own intricacies of which the many controversies in past and recent literature are eloquent witnesses. In order to prevent long-standing friendships being spoiled, we decided to organize a workshop on the renormalization of quantum field theories with non-linear field transformations. It was held at Ringberg Castle in Upper Bavaria, which is owned by the Max Planck Society, from February 16 to 20, 1987. Our intention was to provide a forum for discussions between the specialists and proponents in the field with the hope of producing a fair description of the problems and results and also to achieve some consensus about which questions are resolved to everybody's satisfaction and which may still be considered to require further study.

Part I serves as an introduction. It contains first a short exposition of general renormalization theory, followed by some examples (in four dimensions) where non-linear transformations of fields have been successfully mastered: BRS transformations, those occurring in supersymmetric theories and those in the case of radiative mass generation.

Part II deals with the main topic: the construction of non-linear σ-models. Our hope is that all important branches of this vast subject are represented. It was here that the discussion session was lively and fruitful, hopefully resulting in the above-mentioned consensus.

In Part III the relations of non-linear σ-models to string theory are touched upon, though lack of time prevented a more thorough treatment of this exciting area of research.

Although judgement as to whether our plans were successful should be left to the participants in the meeting and to the readers of these proceedings, we would like to acknowledge the optimal working conditions provided by the locality and the staff of Ringberg Castle on behalf of all the participants. We are indebted to the Max Planck Society for having given us the opportunity to organize this workshop.

Munich
February, 1988

P. Breitenlohner
D. Maison
K. Sibold

Table of Contents

Part I

Non-linear Field Transformations in 4 Dimensions

Transforming fields non-linearly causes problems in quantum field theory: products of fields at one and the same space-time point are singular and hence have to be made well-defined prior to any application. The ambiguities inherent in any such renormalization have to be understood and to be taken care of.

These remarks constitute the program for the present part: First it is recalled what renormalization is about; then those examples are presented where non-linear field transformations have been mastered (in 4-dimensional space-time).

Renormalization Theory, a Short Account of Results and Problems*

DIETER MAISON

Max-Planck-Institut für Physik und Astrophysik
– Werner-Heisenberg-Institut für Physik –
P.O.Box 40 12 12, Munich (Fed. Rep. Germany)

1. History

Historically Quantum Field theory arose from the attempt to quantize charged particles coupled to the electromagnetic radiation field. Already the first calculations by Dirac, Heisenberg and Pauli treating the interaction between the particles and the radiation field as a small perturbation were plagued by infinities for energies, polarizabilities etc. . Not all of these came as a surprise since infinite self-energies resp. -stresses were already known from the classical theory of point particles coupled to the electromagnetic field. Although it was remarked that from a pragmatic point of view the parameters of non-interacting (bare) particles or fields are unobservable and can therefore be made suitably infinite in order to cancel the infinities arising from the interaction, this position is quite unsatisfactory as it renders the starting point of the calculations, the Lagrangean, ill-defined. A more satisfactory attitude is to take the divergencies as an indication that the theory is incomplete and should be embedded in a theory behaving more decently at short distances resp. at large momenta. A divergent but renormalizable theory could then be considered as an 'effective' low energy approximation which is made self-consistent by the renormalization of a finite number of parameters diverging with the high energy cut-off.

---

*Chapters 2 and 3 have been added for the convenience of readers less familiar with the formalism of perturbation theory.

In fact we may even learn some interesting things studying the cut-off dependence of the theory considering it as an 'effective' low-energy theory. For instance the question of 'naturalness' of super-renormalizable couplings resp. mass terms arises precisely from the self-consistency of the 'effective' theory.

However, quite independently of the particular 'philosophy' favoured to cope with the undesirable presence of the divergencies, it turns out to be possible to develop calculational procedures avoiding the infinities and reducing them to an ambiguity which can be removed through a fit to the observed values of the parameters ('Renormalization Theory').

In the early days of perturbative quantization the main emphasis was put on finding simple calculational schemes mitigating the unwanted divergencies. However it was soon recognized that subtracting infinity from infinity was not a terribly unique recipe. 'Hence there was a definite need for a structural investigation of the divergencies of QED and its consistent removal' (Dyson). In addition, beyond the one-loop approximation one was faced with a principal problem in form of the so-called overlapping divergencies. The consistent removal of these turned out to be a rather tricky entertainment leading to a satisfactory answer only after a number of erroneous steps about which A. Wightman commented: 'Renormalization Theory has a history of egregious errors by distinguished savants. It has a justified reputation of perversity; a method that works up to 13th order in the perturbation series fails in the 14th order.' Here Wightman refers to a method of Ward to renormalize QED that works perfectly well until one meets graphs of the type

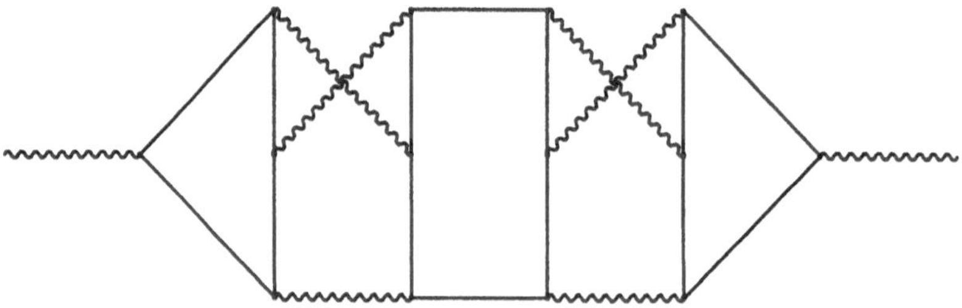

when things go wrong.

The difficulties pertaining to the proper treatment of overlapping divergencies were finally resolved by a systematic approach based on general postulates like locality, unitarity and Poincaré invariance. This 'axiomatic' approach emerging from ideas of Stueckelberg was fully formalized by Bogoliubov and resulted in a rigorous construction of the renormalized perturbation expansion to all orders due to the penetrating work of Hepp. A particularly powerful formulation was given by Zimmermann, who succeeded to resolve the result of the recursive addition of counter-terms to the Lagrangean resp. subtractions of vertex functions on Feynman amplitudes into a closed expression called the 'forest formula'. Many of the further developments of renormalization theory used this particularly lucid formalism.

Characterizing the renormalized theory by abstract principles instead of defining it through a particular subtraction scheme has the advantage that one can study its properties in a scheme independent way. It only remains to show that there exists some method leading to the desired result, whatever method is used in any particular case turns into a matter of convenience. Some renormalization schemes, as e.g. Zimmermann's have simple formal properties making them ideally suited for general considerations, whereas others like dimensional renormalization are more suitable for actual calculations.

An approach staying as closely as possible to the 'axioms' of renormalization theory was given by Epstein and Glaser [1]. Using the x-space support properties of advanced and retarded Green functions they can avoid undefined quantities altogether. The recursive construction of the perturbation series is reduced to the problem of 'cutting' distributions. At this point the usual ambiguities of the result emerge, which can be removed as usual by suitable normalization conditions.

The 'axiomatic' approach also turns out to be a fiducial guide on the treacherous field of theories with local invariance groups. As in the early days of renormalization theory also in this case the situation was and still is characterized by misinterpreted calculational results and inconsistent assumptions leading often to paradoxical conclusions. What we can, however, learn from an excursion into the history of perturbative renormalization is the fact that - as frequently in science - progress is stimulated by these paradoxical results which can only be resolved by clarifying the basic physical requirements masked by complicated calculational procedures erroneously taken to be a substitute for the latter. Clearly that does not mean that we should underestimate the value of intelligent calculational methods which after all make the renormalized perturbation expansion more than an exercise in mathematics. The overwhelming success of perturbative QED in cases like the higher order corrections to the anomalous magnetic moment of the electron or the muon is an impressive example. In fact, for calculations beyond one loop in the Weinberg-Salam theory it may be vital to find a renormalization scheme minimizing the calculational effort exhausting easily the capacities of even the biggest existing computers.

The renormalized perturbation expansion has also been a powerful guide for non-perturbative considerations. Much of the work of LSZ on quantum field theory has been abstracted from the perturbative series. Of particular importance is the development of 'Constructive QFT' emerging from the attempt to use renormalizations as suggested by perturbation theory, but otherwise proceed non-perturbatively. Its recent development is strongly influenced by the close connection between 'euclideanized' relativistic quantum field theories and and the theory of phase transitions in statistical mechanics. The essential conceptual tool is the 'renormalization group' of Wilson, which also provides a new understanding of the concepts of renormalization theory. Renormalizable theories turn out to be related to the fixed points of the renormalization group transformation. This viewpoint supersedes the conventional perturbative classification and may also allow consistent theories which are perturbatively non-renormalizable.

2. The Free Field

The n-dimensional scalar free field $\varphi(x)$ of mass $m \geq 0$ is a Wightman field [2] acting on a Hilbert space of free particles, the Fock space $\mathcal{F}$. $\mathcal{F}$ has the structure of a direct sum $\mathcal{F} = \bigoplus_{N=0}^{\infty} \mathcal{F}_N$ of N-particle spaces $\mathcal{F}_N$ which are symmetric tensor products of the one-particle space $\mathcal{F}_1 = L_2(d\mu)$ with $d\mu = \delta(p^2 - m^2)\Theta(p^0)d^n p$. $\mathcal{F}_0 = \mathbf{C}\Omega$ is the (no-particle) vacuum sector. $\mathcal{F}_0$ and $\mathcal{F}_1$ carry irreducible unitary representations of the Poincaré group through

$$U(\Lambda, a)\Omega = \Omega$$

$$U(\Lambda, a)\Phi(p) = e^{ipa}\Phi(\Lambda^{-1}p)$$

inducing a unitary representation on $\mathcal{F}$.

The free field $\varphi(x)$ may be defined through its truncated Wightman functions [2]

$$w_2^T(x_1, x_2) = (\Omega, \varphi(x_1)\varphi(x_2)\Omega) = \frac{1}{(2\pi)^{n-1}} \int e^{ip(x_1-x_2)} d\mu(p) = i\Delta^+(x_1 - x_2, m^2)$$

$$w_k^T = 0 \quad \text{for} \quad k \neq 2$$

It obeys the field equation

$$(\partial^2 + m^2)\varphi(x) = 0$$

derived from the Lagrangean

$$\mathcal{L}_0 = -\frac{1}{2} \int \varphi(x)(\partial^2 + m^2)\varphi(x) d^n x$$

Similarly one may define a *generalized free field* $\varphi_\rho(x)$ replacing the measure $d\mu(p)$ by a superposition

$$d\mu_\rho(p) = \int \delta(p^2 - \kappa^2)\Theta(p^0) d\rho(\kappa^2)$$

with some (signed) measure $d\rho(\kappa^2)$ leading to the two-point function

$$(\Omega, \varphi_\rho(x)\varphi_\rho(0)\Omega) = i \int \Delta^+(x, \kappa^2) d\rho(\kappa^2)$$

If the moments $K_j = \int \kappa^{2j} d\rho(\kappa^2)$ vanish for $0 \leq j \leq J$ (for J sufficiently big) the two-point function of $\varphi_\rho(x)$ becomes differentiable. Hence generalized free fields $\varphi_\rho(x)$ may be used as regularized versions of $\varphi(x)$.

3. Wick Products

The *Wick products* $:\varphi(X): = :\varphi(x_1)\ldots\varphi(x_k):$ can again be conveniently defined through their vacuum expectation values

$$(\Omega, :\varphi(X_1):\ldots:\varphi(X_k):\Omega) = \sum_{\bigoplus Z_i = \bigoplus X_j} \prod_i w_2^T(Z_i)$$

where the sum runs over all possible ordered pairs $Z_i = (x_{\alpha_1}, x_{\alpha_2})$ where the x_{α_i} are elements of different X_j's.

Example:

$$\begin{aligned}&(\Omega, :\varphi(x_1)\varphi(x_2)::\varphi(x_3)\varphi(x_4):\Omega) = \\ &-\Delta^+(x_1 - x_3)\Delta^+(x_2 - x_4) - \Delta^+(x_1 - x_4)\Delta^+(x_2 - x_3)\end{aligned}$$

The Wick product $:\varphi(X):$ remains well-defined even if all the elements of $X = \{x_1, \ldots, x_r\}$ coincide, leading to the Wick power $\frac{:\varphi^r:(x)}{r!}$.

The definition of Wick products can be generalized to derivatives of $\varphi(x)$ introducing a suitable multi-index notation [1]

$$\frac{:\varphi^r:(x)}{r!} \equiv : \prod_{\alpha \in \mathbf{N}^n} \frac{1}{r(\alpha)!}(\partial^\alpha \varphi(x))^{r(\alpha)}:$$

where only a finite number of $r(\alpha)$ are different from zero.

The vacuum expectation values $(\Omega, \frac{:\varphi^{r_1}:(x_1)}{r_1!} \ldots \frac{:\varphi^{r_k}:(x_k)}{r_k!}\Omega)$ can be evaluated with the formula given above. To each term $\Delta^+(Z_1)\ldots\Delta^+(Z_k)$ corresponds a graph G whose vertices are the x_i and whose lines connect the vertices given by the Z_j's.

Example:

$\Delta^+(x_1 - x_2)\Delta^+(x_1 - x_3)\Delta^+(x_2 - x_3)\Delta^+(x_2 - x_3)$ contributing to

$$(\Omega, \frac{:\varphi^2:(x_1)}{2!}\frac{:\varphi^3:(x_2)}{3!}\frac{:\varphi^3:(x_3)}{3!}\Omega)$$

gives the graph

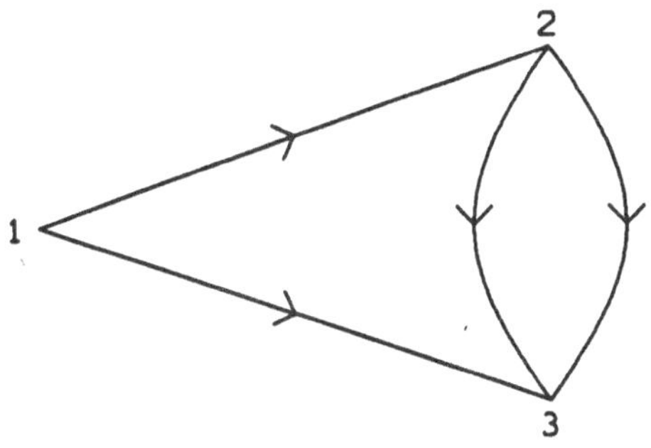

Wick's theorem allows to expand multiple Wick products into simple ones:

$$:\varphi(X_1):\ldots:\varphi(X_k): = \sum_{Y_i \subset X_i} (\Omega, :\varphi(X_1 \setminus Y_1):\ldots:\varphi(X_k \setminus Y_k):\Omega) :\varphi(Y_1)\ldots\varphi(Y_k):$$

From this formula one easily derives the corresponding expansion for products of Wick powers

$$\frac{:\varphi^{r_1}:(x_1)}{r_1!}\ldots\frac{:\varphi^{r_k}:(x_k)}{r_k!} = \sum_{s_i}(\Omega, \frac{:\varphi^{(r_1-s_1)}:(x_1)}{(r_1-s_1)!}\ldots\frac{:\varphi^{(r_k-s_k)}:(x_k)}{(r_k-s_k)!}\Omega):\frac{\varphi^{s_1}(x_1)}{s_1!}\ldots\frac{\varphi^{s_k}(x_k)}{s_k!}:$$

Analogous formulae hold for generalized free fields. Sufficiently regularized free fields $\varphi_\rho(x)$ allow for the definition of the *time-ordered functions* resp. *products* $(\Omega, T:\varphi_\rho(X_1):\ldots:\varphi_\rho(X_k):\Omega)$ obtained by replacing the Δ^+ functions by (regularized) *Feynman propagators*

$$\Delta_{F,\rho} = \frac{1}{(2\pi)^n}\int \frac{e^{ipx}}{p^2-\kappa^2+i0} d^n p\, d\rho(\kappa^2)$$

The corresponding graphs are called *Feynman graphs*.

The generalization of Wick's theorem to time-ordered products (well-defined only for regularized fields) is

$$T:\varphi(X_1):\ldots:\varphi(X_k): = \sum_{Y_i \subset X_i} (\Omega, T:\varphi(X_1 \setminus Y_1):\ldots:\varphi(X_k \setminus Y_k):\Omega) :\varphi(Y_1)\ldots\varphi(Y_k):$$

4. The Scattering operator

The scattering operator S (S-matrix) providing a unitary map between the Fock spaces of in- and outgoing asymptotic particles can be characterized by 'axioms' derived from its physical interpretation. Following Bogoliubov [3] one considers the scattering operator $S(g)$ in the presence of 'external' classical fields $g(x)$ assumed to be smooth and localized (e.g. of compact support) which are coupled to suitable quantum fields. The corresponding interacting quantum fields can then be defined by

$$\mathcal{O}(x) = S(g,0)^{-1} \frac{\delta}{i\delta h(x)} S(g,h)|_{h=0}$$

where we have distinguished the particular field $\mathcal{O}$ by its external field $h(x)$.

In order to avoid problems with interactions of infinite duration resp. spatial extension it is convenient to replace also the coupling constants by such localized functions. The *adiabatic limit* $g(x) \to$ const. can then be studied separately. Hence we shall for the moment not distinguish between external fields and coupling constants.

The required properties of $S(g)$ are:

i) $S(0) = \mathbf{1}$ (Normalization)

ii) $U(\Lambda,a)S(g)U(\Lambda,a)^{-1} = S(D(\Lambda)g(\Lambda^{-1}(x-a)))$ (Poincaré invariance) where $D(\Lambda)$ is the finite dimensional representation of the homogeneous Lorentz group corresponding to the covariance of $g(x)$

iii) $S(g)S^{+}(g) = S^{+}(g)S(g) = \mathbf{1}$ (Unitarity)

iv) If the support of g lies outside the causal past of the support of h, i.e. supp $g \cap (\text{supp h} + \bar{V}^{-}) = \emptyset$, then $S(g)^{-1}S(g+h) = S(h)$ (Causality)

The perturbation expansion of $S(g)$ is a power series in the coupling constants resp. external fields $g(x)$

$$S(g) = \mathbf{1} + \sum_{k=1}^{\infty} \frac{i^k}{k!} \int T_k(x_1,\ldots,x_k) g(x_1)\ldots g(x_k) \prod dx_i$$

In order to avoid questions of convergence of the series it is usually interpreted as a formal power series in g. Since the individual terms of the expansion are in general unbounded operators some care has to be taken to find a suitable common invariant domain in $\mathcal{F}$ [1]. As long as one studies only Green functions (compare next paragraph) this problem is avoided.

Given the Lagrangean $\mathcal{L} = \mathcal{L}_0 + \mathcal{L}_{int}$ one can use the canonical formalism to derive a formal expression for $S(g)$, the so-called *Gell-Mann Low formula*

$$\begin{aligned} S(g) &= T\exp\Big(\frac{i}{\hbar}\int \mathcal{L}_{int}(x)dx\Big) \\ &= \sum \frac{1}{k!}\Big(\frac{i}{\hbar}\Big)^k \int T\mathcal{L}_{int}(x_1)\dots\mathcal{L}_{int}(x_k)\prod dx_i \end{aligned}$$

where the symbol T denotes time ordering.

From this expression for $S(g)$ one reads off $T_1(x)g(x) = \frac{1}{\hbar}\mathcal{L}_{int}(x)$.

For the definition of the interacting field tending asymptotically to $\varphi(x)$ it is necessary to include in $\mathcal{L}_{int}$ a term $\int \varphi(x)j(x)\,d^nx$ linear in the free field φ. This implies [4] that $\mathcal{L}_{int}$ must be a *finite* linear superposition of Wick monomials $\frac{:\varphi^r:(x)}{r!}$. (For dimension $n = 2$ the free field $\varphi(x)$ has vanishing canonical dimension; in this case also infinite sums are possible.)

In contrast to the first non-trivial term in the Gell-Mann Low formula all the higher ones are in general ill-defined due to the T-products involved. A procedure constructing well-defined T_n's to a given T_1 resp. $\mathcal{L}_{int}$ in accordance with the 'axioms' i)-iv) is called a *renormalization.* Such renormalizations involve in general some arbitrariness. This arbitrariness is however severely restricted by the validity of the axioms i)-iv) as expressed by the

THEOREM A [3]:

Two renormalized perturbation expansions for $S(g)$ fulfilling the axioms i)-iv) which coincide up to the $(k-1)^{th}$ order $(k \geq 2)$ may differ at the kth order at most by a completely local term of the form

$$\Lambda_k(x_1,\dots,x_k) = \sum_r P_r(\partial)\delta(x_1-x_2)\dots\delta(x_{k-1}-x_k)\frac{:\varphi^r:(x_1)}{r!} \qquad (*)$$

where the P's are differential operators further restricted by conditions ii) and iii).

Such a difference is called a *finite renormalization.* Another central result of renormalization theory is that such finite renormalizations can be absorbed into a redefinition of T_1 i.e. $\mathcal{L}_{int}$.

THEOREM B [3]:

Given two renormalized expansions $S^{(1)}(g)$ and $S^{(2)}(g)$ for the same $\mathcal{L}_{int}$ one can find a set of $\Delta\mathcal{L}_k$'s of the form (*) such that after the replacement

$$\mathcal{L}_{int}(x) \to \mathcal{L}_{int}(x) + \sum_{k\geq 2} \frac{1}{k!} \int \Delta\mathcal{L}_k(x, x_1, \ldots, x_{k-1}) \prod dx_i$$

in the construction of say $S^{(1)}(g)$ they coincide.

The local operators $\Delta\mathcal{L}_k(x_1, \ldots, x_k)$ can be constructed recursively from the Λ_n's:

$$\Delta\mathcal{L}_1 = \Lambda_1 \qquad \Delta\mathcal{L}_2 = \Lambda_2 - \frac{2i}{\hbar} T\mathcal{L}_{int}\Lambda_1 \quad \text{etc.}$$

In order to make the construction of $S(g)$ independent of the renormalization scheme it is necessary to specify a set of *normalization conditions* (compare below).

An important corollary to Theorem B is the relation between two different renormalizations of the same (composite) interacting quantum field, distinguished by its source $h(x)$, i.e. $\mathcal{O}(x) = S(g,0)^{-1} \frac{\delta}{\delta h} S(g,h)|_{h=0}$

COROLLARY:

The interacting quantum fields $\mathcal{O}^i(x)$ $(i = 1, 2)$ referring to two different renormalized S-matrices $S^{(i)}$ for the same $\mathcal{L}_{int}$ such that $S^{(1)}(g,0) = S^{(2)}(g,0)$ are related by the *Zimmermann identity* [5]

$$\mathcal{O}^{(1)} - \mathcal{O}^{(2)} = \sum_j r_j(g) \mathcal{O}_j^{(1)}(x)$$

where the sum runs over a suitable basis of interacting quantum fields $\mathcal{O}_j$ and the coefficients $r_j(g)$ are at least of order g.

Theorem B is also exploited in the renormalization through 'infinite' counterterms added to $\mathcal{L}_{int}$. Introducing regularized (cut-off) fields φ_ρ one may take the naive Gell-Mann Low formula for $S(g)_\rho$ replacing $\mathcal{L}_{int}$ by $\mathcal{L}_{int} + \Delta\mathcal{L}_\rho$ where $\Delta\mathcal{L}_\rho$ is chosen such that $S(g)_\rho$ has a well-defined limit with the properties i)-iv) when the regularization ρ is removed.

If the original Lagrangean contains all the terms required as counterterms the replacement $\mathcal{L} \to \mathcal{L} + \Delta\mathcal{L}$ amounts to a replacement $g \to g_{ren} + \Delta g_\rho$ and a renormalization of $\mathcal{L}_0$. In order to decide which terms are required as counterterms or, to pose it differently, what is the minimal ambiguity introduced through the renormalization process one has to control the singularity of $T_n(x_1, \ldots, x_n)$ for coinciding arguments. This is achieved through the so-called *Power Counting Rules* based on the assignment of a canonical dimension d to Wick powers

$$d(\frac{:\varphi^r:}{r!}) = \sum_\alpha (\frac{n-2}{2} + |\alpha|)\, r(\alpha)$$

From this one arrives at the *UV power counting degree* ω (degree of UV singularity) of the Green functions $(\Omega, T\frac{:\varphi^{r_1}:(x_1)}{r_1!} \ldots \frac{:\varphi^{r_k}:(x_k)}{r_k!}\Omega)$ considered as distributions [1].

$$\omega(r_1, \ldots, r_k) = \sum_i (d(\frac{:\varphi^{r_i}:}{r_i!}) - n) + n$$

To the time ordered product $T\frac{:\varphi^{r_1}:(x_1)}{r_1!} \ldots \frac{:\varphi^{r_k}:(x_k)}{r_k!}$ one assigns the degree of its vacuum expectation value.

As observed above $\mathcal{L}_{int}$ is (for $n \neq 2$) a finite sum of Wick monomials. In this case this power counting formula allows to restrict the ambiguity in the construction of the T_k's resp. to characterize the type of counter terms required in $\Delta\mathcal{L}_k$. Theories for which the degree ω of the counterterms $\Delta\mathcal{L}_k$ does not increase with k are called *renormalizable* (by power counting). For this classification one takes into account only those terms of $\mathcal{L}_{int}$ referring to the genuine coupling constants. Among the renormalizable theories one distinguishes *super-renormalizable* theories requiring only a finite number of counterterms, i.e. $\omega \geq 0$ only for finitely many T_k's, and *strictly renormalizable* theories for which the degree of the Wick polynomial $\Delta\mathcal{L}_k$ is independent of k. Theories for which the degree of $\Delta\mathcal{L}_k$ grows with k and which therefore involve necessarily infinitely many coupling constants are called *non-renormalizable*. For $n = 2$ or more generally for fields of vanishing canonical dimension some other concept of renormalizability is required.

Obviously power counting also restricts the coefficients in the Zimmermann identity.

5. Generating Functions

The generating functional

$$Z(j) = \sum_{k=0}^{\infty} \frac{i^k}{k!} \int \tau^{(k)}(x_1, \ldots, x_k) j(x_1) \ldots j(x_k) dx_1 \ldots dx_k$$

of the time-ordered Green functions is given in terms of the S-matrix by $Z(j) = (\Omega, S(g,j)\Omega)$, where we have distinguished the external field coupling to $\varphi(x)$ by the letter j. From the knowledge of $Z(j)$ one can completely reconstruct the operator $S(g,j)$. The perturbation expansion of the Green functions yields terms of the type

$$(\Omega, T\varphi(x_1) \ldots \varphi(x_k) \frac{:\varphi^{r_1}:(y_1)}{r_1!} \ldots \frac{:\varphi^{r_l}:(y_l)}{r_l!} \Omega)$$

which can be represented by Feynman graphs.

The generating functional for *connected* Green functions is given by $Z_c(j) = \ln Z(j)$. This terminology refers to the fact that connected Green functions correspond to connected Feynman graphs.

For the renormalization the *generalized vertices* or *one-particle irreducible* (1PI) Green functions are most important, because their response to a change of $\mathcal{L}_{int}$ is most transparent and in addition their renormalization is sufficient to make $Z(j)$ well-defined [3]. Their generating functional $\Gamma(\phi)$ is obtained from $Z_c(j)$ through a Legendre transformation. Setting

$$\phi(x) \equiv \frac{\delta}{\delta j(x)} Z_c(j) - \phi_0$$

with $\phi_0 = \frac{\delta}{\delta j(x)} Z_c(j)\Big|_{j=0}$ and resolving this equation (recursively) with respect to $j(x)$ as a functional of $\phi(x)$ one puts

$$\Gamma(\phi) = Z_c(j(\phi)) - \int j(x)\big(\phi(x) + \phi_0\big) dx$$

To the order g the vertex functional $\Gamma(\phi) = \mathcal{L}_{int}$ where on the r.h.s. the quantum field $\varphi(x)$ is replaced by $\phi(x)$.

Frequently $\Gamma(\phi)$ is considered as a formal power series in $\hbar$ whose powers count the number of loops of the corresponding Feynman graphs.

The power counting rules allow to assign a degree ω to each 1PI Green function $\Gamma^{(k)}$ in the expansion

$$\Gamma(\phi) = \sum_{k=0}^{\infty} \frac{i^k}{k!} \int \Gamma^{(k)}(x_1, \ldots, x_k)\phi(x_1)\ldots\phi(x_k)dx_1\ldots dx_k$$

The ambiguity in the construction of $S(g)$ can be removed in a transparent way posing $\omega + 1$ normalization conditions for all $\Gamma^{(k)}$'s with $\omega \geq 0$.

Example: Taking $\int \mathcal{L}_{int}(x)d^4x = g\int \frac{:\varphi^4:(x)}{4!}d^4x$ and $m^2 > 0$ one may pose the normalization conditions in momentum space

$$\tilde{\Gamma}^{(2)}(p,-p)|_{p^2=m^2} = 0$$

$$\frac{\partial}{\partial p^2}\tilde{\Gamma}^{(2)}(p,-p)|_{p^2=m^2} = i$$

$$\tilde{\Gamma}^{(4)}(p_1,p_2,p_3,p_4)|_{p_i^2=m^2,(p_i+p_j)^2=\frac{4}{3}m^2(i\neq j)} = -ig$$

where we have suppressed δ-functions expressing momentum conservation and assumed that the symmetry $\varphi \to -\varphi$ is preserved.

6. Infrared Problems

Infrared problems may arise in theories involving massless fields in the 'adiabatic limit', i.e. when the space-time cut-offs from the genuine interaction Lagrangean are removed. In the case $m \neq 0$ it can be proven [6] that the *strong adiabatic limit* for $S(g)$ exists in $\mathcal{F}$ if suitable normalization conditions are properly taken into account. On the other hand, if massless fields are involved the strong adiabatic limit will in general not exist, because the asymptotic fields are not really free i.e. non-interacting. In order to avoid this 'dynamical' IR problem one may however study the *weak adiabatic limit* of the off-shell Green functions $\tau^{(k)}$ resp. $\Gamma^{(k)}$. At this level incurable (perturbative) IR deseases occur, if massless fields have super-renormalizable couplings. In order to control the situation suitable *IR power counting rules* were developed [7, 8].

But even when the result of IR power counting is admissible it is still necessary to guarantee that massless fields do not develop super-renormalizable couplings through radiative corrections. This has to be insured through proper normalization conditions for certain vertex functions $\Gamma^{(k)}$ of these fields. In certain cases this may be impossible due to an IR-instability of the interaction Lagrangean. If this happens it is necessary to take these radiative super-renormalizable couplings into account already in the tree approximation thus changing completely the perturbation expansion [9].

Another form of such off-shell IR problems arises with fields of canonical dimension zero. A typical case are canonical scalars in two dimensions, e.g. the coordinate fields of non-linear σ-models. A further case are sub-canonical scalars as e.g. the lowest component of vector superfields in four dimensions [10]. In these cases already the free propagator needs an IR regulator. Only very special Green functions have a chance to have a decent adiabatic limit independent of the IR regulator in this case. In addition this perturbative limit may not give the correct physical result due to an illegal interchange of limits (compare discussion session !).

7. The Action Principle

Schwinger's action principle [11] first studied by Lowenstein [12] and Lam [13] in its renormalized form describes the change of the Green functions under infinitesimal variations of the Lagrangean. It is particularly important in the study of symmetry transformations acting on the quantized fields.

In its naive form the action principle has two parts:

i) Infinitesimal variations of external fields, coupling constants (treated on the same footing in this setting) resp. parameters in $\mathcal{L}_0$ result in

$$\frac{\delta}{\delta g(x)} Z(g) = (\Omega, T \frac{\delta}{\delta g(x)} \int \mathcal{L}(x)\, d^n x\, e^{\frac{i}{\hbar}\int \mathcal{L}_{int}(x)\, d^n x}\Omega) \, .$$

The insertion of $\frac{\delta}{\delta g(x)} \int \mathcal{L}(x)\, d^n x$ inside the time-ordered product is called an *operator insertion* [12] and also denoted by $\frac{\delta}{\delta g(x)} \int \mathcal{L}(x)\, d^n x\, Z(g)$.

ii) Infinitesimal variations of the quantized field φ yield

$$\delta Z(g) = (\Omega, T \frac{\delta \mathcal{L}}{\delta \varphi} \delta\varphi(x)\, e^{\frac{i}{\hbar}\int \mathcal{L}_{int}(x)\, d^n x}\Omega) = 0 \, .$$

For the generating functional of the renormalized Green functions $Z(g)$ the insertions calculated naively pick up radiative corrections. So for example ii) is changed into

$$\delta Z(g) = (\Omega, T \delta\mathcal{L}(x)\, e^{\frac{i}{\hbar}\int \mathcal{L}_{int}(x)\, d^n x}\Omega) = \Delta Z(g)$$

where Δ is a local operator insertion which is at least of the order $\hbar$.

8. Symmetries

Up to now we have discussed the renormalization of theories for which the Lagrangean contains a complete set of Lorentz covariant Wick monomials of a scalar field compatible with the power counting rules. In practice one is, however, interested in theories implementing special global or local symmetry transformations. At the classical level (tree approximation) this requires the introduction of multiplets of fields with possibly different Lorentz covariance properties and the construction of invariant Lagrangeans. Upon quantization it is important to give precise conditions how such symmetries are to be implemented since the Lagrangean as well as possibly non-linear field transformations are no more well defined. This remark is also to be understood as a further *warning* that it is not enough to take some classical Lagrangean, write down Feynman rules and subtract infinities. In view of the ambiguities inherent in the renormalization procedure it is important to give conditions on the renormalized generating functional $Z(g)$ resp. $\Gamma(\varphi)$ expressing the symmetry of the theory and making the result renormalization scheme independent.

In the case of continuous symmetries it is standard to consider the variation of $Z(g)$ under infinitesimal transformations of the fields. Suppose we have a multiplet of fields $\Phi(x)$ (elementary or composite) closed under the local infinitesimal transformations

$$\delta_i\Phi(x) = t_i\Phi(x)$$

with some constant matrices t_i and that the (classical) Lagrangean has the structure

$$\mathcal{L}(x) = \mathcal{L}_{inv}(x) + J^{\mathrm{T}}(x)\Phi(x)$$

then the classical action $\int \mathcal{L}dx$ satisfies

$$(\delta_i + W_i)\int \mathcal{L}dx = 0$$

where the W_i's are differential operators

$$W_i = -\int J^{\mathrm{T}}(x)t_i\frac{\delta}{\delta\Phi^{\mathrm{T}}}(x)dx$$

The naive action principle ii) implies the Ward identities

$$W_iZ(g,J) = 0$$

We may now require the same Ward identity for the renormalized generating functional as a substitute for the invariance of the quantized Lagrangean.

There are essentially two different strategies to construct renormalized Green functions obeying the Ward identities:

i) take a manifestly invariant renormalization scheme, e.g. using some invariant regularization;

ii) take an arbitrary renormalization scheme and exploit the freedom to perform finite renormalizations (compatible with power counting) to enforce the validity of the Ward identities.

It turns out that there are cases, where even strategy ii) fails (e.g. Adler-Bardeen anomaly). Using the Lie algebra structure of the differential operators W_i it is possible to characterize possible obstructions (called *anomalies*) to the construction of a $Z(g, J)$ fulfilling the Ward identities [14]. The study of these anomalies gives rise to interesting problems in cohomology theory which triggered a fruitful dialog between mathematicians and physicists.

9. Non-linear field transformations

Particular problems arise with non-linear field transformations typical for fields taking their value on general manifolds (e.g. non-linear σ-models). In view of the geometrical nature of such theories one has to require that also the quantized theories should be invariant under transformations compatible with the geometric structure of the manifold, a.e. diffeomorhisms (general coord. transfs.), affine transfs. (gauge transfs.), isometries (rigid motions) etc.. In order to linearize them as required for Ward identities one is in general forced to introduce an infinite string of composite fields. A simple example is

$$\delta\varphi = \varphi^2 \quad , \quad \delta\varphi^2 = 2\varphi^3 \quad , \ldots$$

A possible way out of this dilemma seems to be the introduction of anti-commuting parameters or ghost fields à la BRS, such that $\delta^2 \equiv 0$ (compare the lectures by Stora and Blasi).

10. Specific Regularization Schemes

Obviously it is very convenient to use an invariant regularization resp. renormalization scheme, if some symmetry is to be implemented, since the number of terms in $\mathcal{L}_{int}$ to be taken into account is in general much larger for a non-invariant procedure. This is a particular problem for local gauge invariances which are highly restrictive.

There are essentially two schemes that have been invented to deal with gauge theories:

i) dimensional regularization;

ii) higher covariant derivative regularization.

Dimensional regularization works well for vector gauge theories like QCD, but fails for chiral gauge theories like the Weinberg-Salam theory. Even when the axial anomaly cancels algebraically there are troubles with γ_5 in the dimensional scheme. Also for supersymmetric theories dimensional regularization is not suitable (in the sense of an invariant regularization). The modification proposed by Siegel [15] known as 'regularization by dimensional reduction' is plagued by inconsistencies whose effects beyond the 1-loop approximation are not under control.

The higher covariant derivative method advertized in [16] does not seem to yield a consistent renormalization procedure (compare the lecture by Sénéor).

In view of this situation it would seem highly desirable to invent some invariant regularization scheme for chiral gauge theories with algebraic anomaly cancellation. This could perhaps terminate the state of confusion about the applicability of the dimensional scheme prevalent in the present literature [17].

REFERENCES

[1] H. Epstein and V. Glaser, *Ann. Inst. Henri Poincaré* **29** (1973) 211.

[2] R. Jost, *The General Theory of Quantized Fields*, Amer. Math. Soc., Providence R.I., 1965.

[3] N.N. Bogoliubov and D.V. Shirkov, *Introduction to the Theory of Quantized Fields*, Wiley-Intersience, New York, 1959.

[4] H. Epstein, *Nuov. Cim.* **27** (1963) 886.

[5] W. Zimmermann, *Ann. Phys. (N.Y.)* **77** (1973) 536.

[6] H. Epstein, in *Renormalization Theory*, G. Velo and A. Wightman eds., Dordrecht, 1976.

[7] J.H. Lowenstein and W. Zimmermann, *Nucl. Phys.* **B 86** (1975) 77.

[8] P. Breitenlohner and D. Maison, *Commun. Math. Phys.* **52** (1977) 55.

[9] G. Bandelloni, C. Becchi, A. Blasi and R. Collina, *Commun. Math. Phys.* **67** (1978) 147.

[10] O. Piguet and K. Sibold, *Nucl. Phys.* **B 247** (1984) 484, *Nucl. Phys.* **B 248** (1984) 336 and *Nucl. Phys.* **B 249** (1984) 396.

[11] J. Schwinger, *Phys. Rev.* **82** (1951) 914, *Phys. Rev.* **91** (1953) 713.

[12] J. Lowenstein, *Commun. Math. Phys.* **24** (1971) 1.

[13] Yuk-Ming P. Lam, *Phys. Rev.* **D 8** (1973) 2943.

[14] C. Becchi, A. Rouet and R. Stora, *Ann. Phys. (N.Y.)* **98** (1976) 287.

[15] W. Siegel, *Phys. Lett.* **84 B** (1979) 193.

[16] L.D. Faddeev and A.A. Slavnov, *Gauge Fields, Introduction to Quantum theory*, Benjamin, Reading, 1980.

[17] R. van Damme, *Nucl. Phys.* **B 227** (1983) 317;
M.E. Machacek and M.T. Vaughn, *Nucl. Phys.* **B 222** (1983) 83;
I. Jack and H. Osborn, preprint DAMTP 84/2.

SOME GENERAL REFERENCES ON RENORMALIZATION THEORY

N.N. Bogoliubov and D.V. Shirkov, *Quantum Fields*, Benjamin, Cummings, Reading Mass., 1983.

E.R. Speer, *Generalized Feynman Amplitudes*, Princeton University Press, Princeton, 1969.

K. Hepp, *Théorie de la renormalisation*, Lecture Notes in Physics Vol. 2, Springer, Berlin, 1969.

C. de Witt and R. Stora eds., *Renormalisation Theory in Statistical Mechanics and Quantum Field Theory*, Gordon and Breach, New York, 1970.

G. Velo and A.S. Wightman eds., *Renormalization Theory*, Reidel, Dordrecht, 1976.

O. Piguet and A. Rouet, *Symmetries in perturbative quantum field theory*, *Phys. Rep.* **76** (1981) 1.

C. Becchi, *The renormalization of gauge theories*, Les Houches 1983, B.S. deWitt and R. Stora eds., Elsevier, 1984.

O. Piguet and K. Sibold, *Renormalized Supersymmetry*, Birkhäuser, Boston, 1986.

SOME REMARKS FOR THE CONSTRUCTION OF YANG-MILLS FIELD THEORIES

Roland Sénéor
Centre de Physique Théorique, Ecole Polytechnique
91128 Palaiseau Cedex, France

One of the most challenging problem that people which are fond of rigorous results in physics would like to solve is to prove the existence of the non-perturbative Yang-Mills model.

I will present here the first step of an approach to this question worked out in collaboration (1) with J. Feldman from U.B.C. (Canada) and J. Magnen and V. Rivasseau from Ecole Polytechnique (France). To precise the goal I will say that our ambition is to construct a finite volume pure Yang-Mills Euclidean field theory in the small coupling regime. We also fix the gauge group to be SU(2), this last restriction being only for notational reason. Adding some matter fields will probably not make the problem much more harder. The only strong hypothesis are the one concerning the finitness of the volume since the removal of this condition will mean that we know how to deal with the infrared problem in gauge theory and the one concerning the smallness of the coupling constant. This will be an alternative to the lattice approach of T. Balaban (2).

In fact in the last few years, we (3) and others (4) were able to construct asymptotically free Euclidean field theories like the Gross-Neveu in 2 dimensions and the infrared φ^4_4 in 4 dimensions. The only difference between Yang-Mills and these theories is the gauge invariance. Up to now the approaches to this problem try not to break the symmetries due to this invariance. Either they were attempts to build fields on some invariant manifolds as the space of orbits (5) or to approximate them in gauge invariant way (6) or, they preserve invariance by looking at the Wilson action on a lattice.

To construct something as complicated as Yang-Mills fields one needs to start from some simpler objects that we know for sure and to get the whole theory as a limiting process defined with their help. In the lattice case, the lattice plays the role of an ultraviolet cutoff and for each lattice size one has through the Wilson action a well defined model (7) the main problem is therefore to go to the continuous limit i.e. to let the lattice size go to zero. In this case one work with the group, the fields which are agebra elements are recovered at the limit. The Federbush approach use the continuum fields are smeared in order to define lattice elements. The purely continuous version tries to define a regularized invariant diffusion process on the invariant manifold approching close enough the Yang-Mills action.

The purpose of this talk is to show that it is possible to start with an approximation which does not preserve gauge invariance, the gauge invariance being recovered when we remove the approximation. In other word the Yang-Mills action is stable with respect to some perturbations which break gauge invariance. It seems surprisingly that this stability was never questioned before.

In a first section we recall what are the main ingredients which are needed to prove the existence of an asymptotically free Euclidean field theory. We then discuss a possible choice of covariant regularization and show that it does not work. Finally we explain what is this stable non covariant way of regularizing the theory.

1. Survey of the methods used in constructive field theory

In the constructive approach to asymptotically free models (such as Gross-Neveu in 2 dimensions or the infrared Φ^4 in 4 dimensions) when dealing with expressions of the form

$$\int e^{-S(\Phi)} \prod_x d\Phi(x)$$

one generally splits the action S into 2 parts L_I and L_o and define a reference measure with the free part L_o: it is a Gaussian measure which is perfectly well defined in the Euclidean framework. Then integral above is replaced by

$$\int e^{-L_I(\Phi)} d\mu(\Phi)$$

For the simplicity of notations we will take as reference model an ultraviolet asymptotically free bosonic scalar field theory.

The basic tools to control the theory are

a) truncated perturbation expansions for the fields which are small

b) the positivity of the interaction to dominate the fields which are large

c) the asymptotic freedom for the convergence purpose

the notion of small and large fields being explained later. By the positivity of the interaction one means the boundedness from below of the Euclіean action.

How do we use a) b) and c)? The answer is by doing a phase space analysis. One defines slices of momenta $\{M_i\}$ $i=1,2,\ldots,$ $M>1$. Associated to these slices one introduces:

1) a field decomposition at each point of space

$$\Phi = \Sigma\, \Phi^i$$

related to the reference measure $d\mu(\Phi)$. In fact $d\mu(\Phi)$ is a Gaussian measure of mean 0 and covariance C (generally $C = (-\Delta + m^2)^{-1}$) and one writes

$$\tilde{C}(p) = \sum_i \tilde{C}^i(p)$$

where $\tilde{C}^i(p) = \tilde{C}(p)\zeta^i(p)$, with $\zeta^i(p)$ a function which localizes p to be roughly in the slice i, i.e. of order M^i, and

$$\sum_i \zeta^i(p) = 1$$

Thus $d\mu = \prod_i d\mu_i$ from which follows the field decomposition.

2) a space cell decomposition

For each scale i one writes R^d as an union $\mathfrak{D}_i$ of disjoint cubes Δ of volume M^{-di} and each space integration related to this scale is decomposed according to $\mathfrak{D}_i$.

3) a spliting of the fields into high and low momentum ones relatively to the scale i

$$\Phi_{h,i} = \sum_{j\geq i} \Phi^j \qquad \text{and} \qquad \Phi_{l,i} = \sum_{j<i} \Phi^j$$

At first approximation the high momentum fields will be the small fields and the low momentum fields will be the large fields.

Then one performs an expansion made of 2 parts:

an horizontal one (a cluster expansion) to test the spatial coupling between distant cubes of a lattice $\mathfrak{D}_i$. If the theory is massive this gives an exponential clustering

a vertical one to test the coupling between momentum slices. If the renormalization has been performed this also gives an exponential decrease.

The theory is then expanded in terms of graphs whose dominant contribution comes from the lowest order ones. These graphs have vertices localized in the cubes of the lattices $\{\mathfrak{D}_i\}$ and lines given the C^i's. The coupling constant renormalization leads at each vertex to replace the initial coupling by a running one whose index is related to highest momentum line hooked to it. To control the flow of this running coupling constant one needs to know the 4-point function to the one and two loops order. More precisely in the case of the second order loops the divergent subdiagrams have only to be renormalized "usefully".

2. How to regularize the Yang-Mills functional

To define a Yang-Mills functional one therefore needs to start from some regularization. We already discuss some of them in the introduction. Another possible one which is extensively used in perturbation theory is the dimensional regularization. Unfortunately we cannot used it in a framework of functional integration since we don't know how to dimensionally interpolate functional spaces. It remains a method advocated first by B. Lee and J. Zinn-Justin (8) and then by L. Faddeev and A. Slavnov (9): the regularization by higher covariant derivatives. Let us discuss it.

The idea is to add to the Yang-Mills Lagrangian $L_{Y.M.} = 1/4\ F_{\mu\nu}F_{\mu\nu}$, written for example in the Landau gauge, a regularizing term $L_R = 1/4\ \Lambda^{-4}\ D^2F_{\mu\nu}D^2F_{\mu\nu}$ where

$$F^a_{\mu\nu} = \partial_\mu A^a_\nu - \partial^\nu A^a_\mu - gC^{abc}A^b_\mu A^c_\nu$$

$$D^2 = D_\mu D_\mu \qquad \text{and} \qquad D^{ab}_\mu = \delta^{ab}\partial_\mu - gC^{acb}A^c_\mu$$

The indices a,b... are related to the gauge group and summation over them is implicit. Since we choose SU(2) as gauge group, C^{abc} is the completely antisymmetric tensor with 3 indices.

One then extract from the whole Lagrangian the quadratic part to define in the usual way a Gaussian measure. The corresponding propagator has for Fourier transform:

$$D^{ab}_{\mu\nu}(k) = \delta^{ab}\left[(\delta_{\mu\nu} - \frac{k_\mu k_\nu}{k^2}) \frac{1}{k^2 + k^6/\Lambda^4} + \lambda \frac{k_\mu k_\nu}{k^2} \frac{1}{k^2 + 1}\right]$$

The interaction terms are sums of monomial of the form $\partial^\beta A^\alpha$ with $\alpha + \beta = 4$, $\alpha > 2$ when they come from F^2 and $\alpha + \beta = 8$, $\alpha > 2$, when they come from D^2FD^2F. As it is well known all the graphs are regularized except the 1-loop ones which need an extra regularization. It was shown (F-S) that a Pauli-Villars regularization does not break gauge invariance at the 1-loop level and therefore can be used to complete the regularization of the theory.

We claim that renormalization breaks the gauge invariance or conversely that if we want to maintain gauge invariance then there are infinities.

At the 1-loop order, as said before, the Pauli-Villars trick regularizes the diagrams built with the vertices of F^2 leading to a new value of g: $g_R = gz_1/z_2$ with the conventional definition of the counterterms $z_1 \ldots$. Similarly there are ghosts counterterms $\tilde{z}_1$ and $\tilde{z}_2$ and the Ward identities imply $z_1\tilde{z}_2 = z_2\tilde{z}_1$. This implies that the covariant derivative D^{ab}_μ which was a function of g becomes a function of g_R. Gauge invariance will mean that the higher covariant term has to be replaced by

$$z_3 D^2(g_R) F_{\mu\nu}(g_R) D^2(g_R) F_{\mu\nu}(g_R)$$

But there are no new divergences introduced by the vertices of L_R, thus the 8-point function need not to be renormalized. On the other hand there are no corrections of the form $(-\Delta)^2 A^2$. This implies that $z_3 = 1$ and g is unchanged in the higher covariant derivative term, thus leading to a breakdown of gauge invariance.

Another argument leading to the same conclusion has been obtained by P. Breitenlohner and D. Maison.

We therefore choose to regularize the Yang-Mills functional by using an usual Euclidean invariant cutoff, thus breaking gauge invariance.

3. The non covariant regularization

In this section we will discuss what are some of the consequences of introducing a non covariant regulator. All we will say concern the 1-loop order. Comments will be given at the end for the next order.

The results are the following ones:

1) because of the non covariance of the regulator apart from the usual gauge invariant counterterms there are 2 purely Euclidean invariant contributions: a mass correction and a term proportional to $(A_\mu A_\mu)^2$

2) the non gauge invariant logarithmic divergences are in fact finite

3) there is a large class of cutoff functions which have the property of not spoiling the positivity of the interaction.

We choose for simplicity to work in the Feynman gauge with A-field propagator

$$D^{ab}_{\mu\nu}(k) = \delta_{ab}\,\delta^{\mu\nu}\,\frac{1}{k^2}\,\zeta(k) = \delta_{ab}\delta^{\mu\nu}\,D(k)$$

with cutoff ζ. The infrared behaviour can be taken account either by introducing a fictitious mass in the propagator or by the introduction of periodic boundary conditions. The ghost propagator has the same form.

Let us compute the term proportional to $(A_\mu A_\mu)^2 = (|A|)^2$. The diagrams which contribute to it are (see Fig. 1)

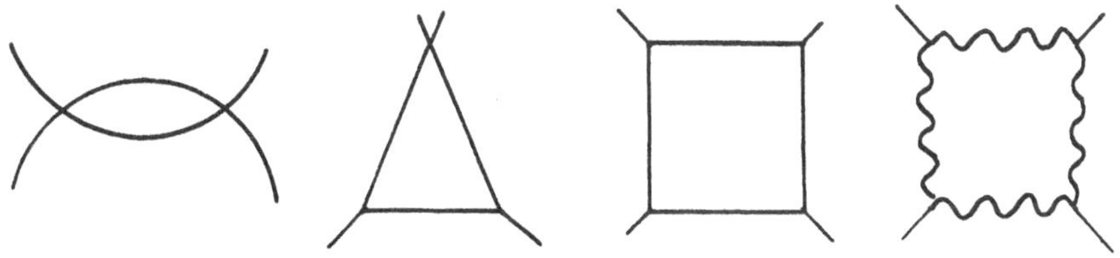

G_1 G_2 G_3 G_4

Fig.1 Contributions to A^4

One finds that the zero momentum contribution is given by

$$36\int (D(k))^2 d^4k \;-\; 90\int k^2 (D(k))^3 d^4k \;+\; 55.5\int (k^2)^2 (D(k))^4 d^4k \;-\; 1.5\int (k^2)^2 (D(k))^4 d^4k$$

As can be seen easily the leading contribution of each integral is Cst $Ln\Lambda$, if Λ is the ultraviolet cutoff and the sum of all the coefficient vanishes; thus the 1-loop contribution to A^4 is finite.

Similarly one can compute the 2-point function $\Gamma^{ab}_{\mu\nu}(p)$. It is proportional to δ^{ab} and one gets

$$\int D(k)D(p-k)[\delta_{\mu\nu}(10(p-k)^2+8k.(p-k)-4(p-k)_\mu(p-k)_\nu-4k_\mu(p-k)_\nu-10k_\mu(p-k)_\nu]d^4k$$

$$-\ 6\delta_{\mu\nu}\int D(k)d^4k\ \ -\ 2\int D(k)D(p-k)[k_\nu(k-p)_\mu]d^4k$$

which correspond respectively to the diagrams Σ_2, Σ_1 and Σ_3 of Fig.2.

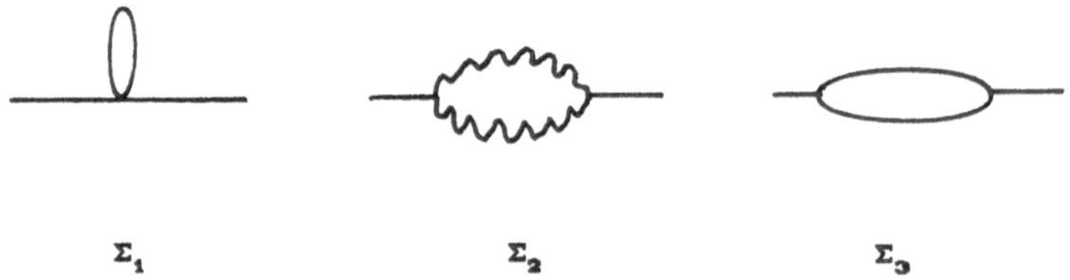

Fig.2 Contributions to Λ^2

One can expand the 2-point function around p=0 and one gets

$$\Gamma^{ab}_{\mu\nu}(p) = \delta^{ab}(\delta_{\mu\nu}b_1 + p_\mu p_\nu b_2 + (p^2\delta_{\mu\nu}-p_\mu p_\nu) + \dots$$

The coefficient b_1 is negative and behaves as Λ^2, but b_2 instead of being logarithmically divergent is finite!

We have thus encounter twice the same mechanism: although we break the gauge invariance the leading logarithmic divergences have gauge invariant coefficients.

At this level, the one loop level, one can understand this in two different ways.

One is by using the proof given in the book of Faddeev and Slavnov that at this order the Pauli-Villars regularization is a gauge invariant regularization. Comparing the effect of the two regularizations one can see easily the above assertion because of the additivity of the logarithms. The other way, which is the one which can be generalized, is by writing Slavnov-Ward identities.

Let us write these identities. One starts with an action of the form:

$$L = -\frac{1}{4}\int\Big[F^2(x) - \frac{\lambda}{2}\mathcal{F}^2(x) - \partial^\mu\bar{\eta}(x)D_\mu\eta(x)\Big]d^4x - \int J^\mu(x)A_\mu(x)d^4x$$

$$-\frac{\lambda}{2}\iint\mathcal{F}(x)(\zeta^{-1}-\delta)(x-y)\mathcal{F}(y)d^4xd^4y + \int\partial^\mu\bar{\eta}(x)(\zeta^{-1}-\delta)(x-y)\partial_\mu\eta(y)d^4xd^4y$$

$$-\frac{1}{4}\iint(\partial^\mu A_\nu-\partial^\nu A_\mu)(x)(\zeta^{-1}-\delta)(x-y)(\partial^\mu A_\nu-\partial^\nu A_\mu)(y)d^4xd^4y$$

The first two integrals are the classical Yang-Mills action and source terms. The other three terms are the corrections due to the introduction of a cutoff. With $\mathcal{F} = \partial^\mu A_\mu$, the quadratic terms give a propagator for the A-field

$$\delta^{ab}\Big[\frac{\delta^{\mu\nu}}{k^2+m^2} - (1-\lambda^{-1})\frac{k^\mu k^\nu}{k^2+m^2}\Big]\zeta(k)$$

Considering the generating functional

$$G(J) = \int e^{L}\prod dA\, d\bar{\eta}d\eta$$

we will obtain some identity by expressing that it is independant of any change of the integration variables. Since a part of L is invariant by the B.R.S. transformation we choose to perform the change

$$A_\mu \longrightarrow A_\mu + \delta A_\mu \qquad \text{with } \delta A^a_\mu = D^{ab}_\mu(x)\eta_b(x)\delta\xi$$

$$\bar{\eta} \longrightarrow \bar{\eta} + \delta\bar{\eta} \qquad \text{with } \delta\bar{\eta}_a(x) = \lambda\,\mathcal{F}_a(x)\delta\xi$$

$$\eta \longrightarrow \eta + \delta\eta \qquad \text{with } \delta\eta_a(x) = -\frac{g}{2}C_{abc}\eta_b(x)\eta_c(x)\delta\xi$$

on

$$I = \int\bar{\eta}_a(x)\, e^{L}\prod dA\, d\bar{\eta}d\eta \equiv 0$$

and look at the first order contribution in $\delta\xi$. We get after an integration by part with respect to the quadratic part and choosing from now on $\lambda = 1$

$$\int\Big[\mathcal{F}_a(x) + \bar{\eta}_a(x)[\int J^\mu_b(u)(D^\mu_{bc}\eta_c)(u)d^4u + g\int C_{bcd}(1-\zeta)(u-v)A^\mu_c(v)\eta_d(v)\frac{\delta}{\delta A^\mu_b(u)}d^4ud^4v$$

$$+\frac{g^2}{2}\iint C_{bcd}C_{bef}(1-\zeta)(u-v)\partial_\mu\bar{\eta}_e(u)A^\mu_f(u)\eta_c(v)\eta_d(v)]\Big]\, e^{L} = 0$$

Introducing the ghost 2-point function $G_{ba}(A;y,x) = \langle\bar{\eta}_a(x)\eta_b(y)\rangle$ one can rewrite it as

$$\int\left[\partial_\mu A^\mu_a(x) + \int J^\mu_b(u)D^\mu_{bc}G_{ca}(A;u,x)d^4u\right] e^{L} =$$

$$= \int\left[- g\int\int C_{bcd}(1-\zeta)(v,u)A^\mu_c(v)\frac{\delta}{\delta A^\mu_b(u)} G_{da}(A;v,x)d^4u d^4v\right.$$

$$\left.- \frac{g^2}{2}\int\int C_{bcd}C_{bef}(1-\zeta)(v,u)A^\mu_f(u)\{G_{ce}(A;v,)\overleftarrow{\delta}_\mu G_{da}(A;v,x) - G_{de}(A;v,u)\overleftarrow{\delta}_\mu G_{ca}(A;v,x)\}\right]e^{L}$$

Expanding G_{ba} as a function of A and g

$$G_{ba}(A;y,x) = -\delta_{ab}\tilde{D}(y-x) - g\int C_{abc}\tilde{D}(y-u)\overleftarrow{\delta}_\mu A^\mu_c(u)\tilde{D}(u-x)\, d^4u$$

$$+ g^2\int\int C_{acd}C_{bce}\tilde{D}(y-v)\overleftarrow{\delta}_\nu\tilde{D}(v-u)\overleftarrow{\delta}_\mu\tilde{D}(u-x)A^\mu_d(u)A^\nu_e(v)\, d^4u d^4v + O(g^3)$$

one now compute to give an example the one loop contribution to the 2-point function. One gets

$$p_\mu\, \Gamma^{\mu\nu}_{ab}(p) = -2\delta_{ab}(p^\nu A(p) - p^2 B^\nu(p)) -4\delta_{ab}C^\nu(p)$$

with

$$A(p) = \int p.k\ D(k)D(p-k)d^4k \quad B^\mu(p) = \int k^\mu\ D(k)D(p-k)d^4k \quad C^\mu(p) = \int(1-\zeta(p-k))k^\mu\ D(k)\ d^4k$$

If we now use the small p expansion of $\Gamma^{\mu\nu}_{ab}(p)$ one sees that

$$p_\mu\Gamma^{\mu\nu} = -4p_\nu C^\nu$$

thus expanding $C^\nu(p)$ as $ap^\nu + bp^2p^\nu + \ldots$ one finds that $b_2 = b$ and

$$b = \frac{3\ \Lambda^{-4}}{2}\int\zeta''(k)\zeta(k)\ d^4k$$

is finite (take ζ of the form $\zeta(k\Lambda^{-1})$).

The general feature which makes the result finite is that the non gauge invariant corrections (which should vanish if $\zeta \to 1$) have in the integrand a contribution in $1 - \zeta$. After a change of variable $k \to k\Lambda^{-1}$ the logarithmic divergence which may appear is cancelled by this difference.

It remains to show that we can choose the cutoff in such a way that the finite contribution to $(|A|)^4$ does not spoil the positivity of the interaction i.e. has a positive sign.

This can be achieved in the following way: take the cutoff function to be of the form $\zeta(k) = 1/2[\eta(\alpha k\Lambda^{-1}) + \eta(\alpha^{-1}k\Lambda^{-1})]$ and use the fact that for any reasonable function η

$$\int \eta(\alpha k\Lambda^{-1})^p \frac{d^4k}{k^2} \approx -\mathrm{Ln}(\alpha\Lambda^{-1}) + \text{finite terms} \qquad \text{if } p\geq 1$$

and for α large enough

$$\int\left[\eta(\alpha k\Lambda^{-1})^p\eta(\alpha^{-1}k\Lambda^{-1})^q\right]\frac{d^4k}{k^2} = \begin{cases} -\mathrm{Ln}(\alpha\Lambda^{-1}) + \text{finite term} & \text{if } p\geq 1 \\ -\mathrm{Ln}(\alpha^{-1}\Lambda^{-1}) + \text{finite term} & \text{if } p=0 \text{ and } q\geq 1 \end{cases}$$

Then the contribution looks like

$$\mathrm{Ln}\alpha \left[-\frac{36\times 2}{4} + \frac{90\times 6}{8} - \frac{54\times 14}{16} \right] + \text{finite terms} = 2.55\ \mathrm{Ln}\alpha + \text{finite terms}$$

where the finite terms are uniformly bounded when Λ and α go to $+\infty$.

Finally it remains to study the 2-loops contributions. The same type of mechanism applies. Again, the main feature is the finitness of non gauge invariant logarithmically divergent terms. Usually 2-loop terms which are logarithmically divergent do not behave as a logarithm but as a square of logarithm. However because we are working in an effective coupling constant scheme it can be shown that the divergence is that of a single logarithm (see (3)) hence leading to the same analysis.

REFERENCES

1 Feldman, J., Magnen, J., Rivasseau, V., Sénéor, R., to be published
2 Balaban, T., Renormalization Group Approach to Lattice Field Theories, Commun. Math. Phys. 109, 249 (1987)
3 Feldman, J., Magnen, J., Rivasseau, V., Sénéor, R., A Renormalizable Field Theory The Massive Gross-Neveu Model in Two Dimensions, Commun. Math. Phys. 103, 67 (1986)
4 Gawedzki, K., Kupiainen, A., Gross-Neveu Model through convergent perturbation expansions. Commun. Math. Phys. 102, 1 (1986)
5 Asorey, M., Mitter, P. K., Regularized Continuum Yang-Mills Process and Feynman-Kac Functional Integral, Commun. Math. Phys. 80, 43 (1981)
6 Federbush, P., A Phase Cell Approach to Yang-Mills Theory, Commun. Math. Phys. 107 319 (1986)
7 Seiler, E., Lecture Notes in Physics, Vol.159, Berlin, Heidelberg, New-York: Springer (1982)
8 Lee, B. W., Zinn-Justin, J., Phys. Rev. D, 5, 3137 (1972)
9 Faddeev, L. D.,Slavnov, A. A., Gauge Fields, Benjamin, Reading (1980)

NON-LINEAR FIELD TRANSFORMATIONS

Simple Examples and General Remarks

K. Sibold
Max-Planck-Institut für Physik und Astrophysik,
- Werner-Heisenberg-Institut für Physik -
Postfach 401212, D-8000 Munich 40, Fed. Rep. of Germany

Table of Contents

1. Linear Symmetry Transformations

In order to familiarize ourselves with the problems to come let us first consider a simple model and perform therein all steps in detail which for more complicated cases will perhaps only be sketched.

Consider an isovector field

$$\underline{\varphi} = \begin{pmatrix} \varphi_1 \\ \varphi_2 \\ \varphi_3 \end{pmatrix} \tag{1}$$

for which an invariant Lagrangian reads

$$\mathcal{L} = \frac{1}{2} \partial\underline{\varphi}\, \partial\underline{\varphi} - \frac{1}{2} m^2 \underline{\varphi}\,\underline{\varphi} - \frac{g}{4!}(\underline{\varphi}\,\underline{\varphi})^2 . \tag{2}$$

The invariance of the classical action

$$\Gamma = \int dx\, \mathcal{L} \tag{3}$$

under the field transformations

$$\delta_i \varphi_j = \varepsilon_{ijk} \varphi_k \tag{4}$$

can be expressed by the Ward-identity (WI)

$$W_k \Gamma \equiv -i \int dx\, \varphi_l\, \varepsilon_{lkm} \frac{\delta\Gamma}{\delta\varphi_m} = 0 . \tag{5}$$

We note in addition that the algebra of the symmetry transformations (4) is translated to the algebra of the WI-operators:

$$[W_k, W_l] = i\,\varepsilon_{klm} W_m \tag{6}$$

i.e. the W's satisfy commutation relations like the angular momentum. The problem to be solved is now easily formulated: extend the classical action to the generating functional for vertex functions

$$\Gamma = \Gamma_{cl} + \hbar\,\Gamma^{(1)} + \hbar^2\,\Gamma^{(2)} + \dots \tag{7}$$

(formal power series), such that the WI (5) holds for Γ. The latter then expresses the symmetry content of the theory.

If we were to use a symmetric regularization and subsequent renormalization we could very simply ensure the existence of (7) and the validity of (5) by giving Feynman rules where φ, m, g are replaced by φ_{ren}, m_{ren}, g_{ren}. Since in general such symmetric regularizations do not exist we treat already the above simple example by a more general method (see R. Stora in [1]).

We invoke the action principle (see Appendix)

$$W_k \Gamma = \Delta_k \cdot \Gamma \tag{8}$$

which tells us that operating with W_k on Γ leads to a local insertion of specified dimension (here 4). This insertion is of order $\hbar$ at least and originates from the fact that our technique of making finite the one-particle-irreducible diagrams was in conflict with the symmetry; (in the classical theory there are no subtractions to be performed, Γ_{cl} was symmetric, whence the order $\hbar$ for Δ_k). Using (cf. Appendix)

$$\Delta_k \cdot \Gamma = \Delta_k + O(\hbar\,\Delta_k), \tag{9}$$

we arrive at

$$W_k \Gamma = \Delta_k + O(\hbar\,\Delta_k) . \tag{10}$$

Up to now the insertion Δ_k was restricted only by locality and power counting, but the algebra (6) implies additional constraints. Acting with W_l on (10), subtracting it from the corresponding relation with l,k interchanged and using (6) we derive the so-called Wess-Zumino consistency conditions

$$W_k \Delta_l - W_l \Delta_k = i\,\varepsilon_{klm} \Delta_m + O(\hbar\,\Delta) . \tag{11}$$

Hence to lowest order of $\hbar$ we have reduced the quantum problem of establishing (5) to solving the classical problem

$$W_k \Delta_l - W_l \Delta_k = i\,\varepsilon_{klm} \Delta_m \tag{12}$$

where Δ_k consists of sums of polynomials in φ of dimension less than or equal to four. Multiplying (12) by ε_{jkl} and summing over k,l we have the condition

$$\Delta_j = -i\, \varepsilon_{jkl}\, W_k \Delta_l \, . \tag{13}$$

In order to solve (13) for Δ_l we use (13) again

$$\begin{aligned} \Delta_j &= -i\, \varepsilon_{jkl}\, W_k \left(-i\, \varepsilon_{lmn}\, W_m \Delta_n \right) \\ &= -(\delta_{jm}\delta_{kn} - \delta_{jn}\delta_{km})\, W_k W_m \Delta_n \\ &= -\, W_n W_j \Delta_n + \underline{W}.\underline{W}\, \Delta_j \, . \end{aligned} \tag{14}$$

We have produced the Casimir operator for our group!

$$\begin{aligned} (1 - \underline{W}\underline{W})\Delta_j &= -\, W_n W_j \Delta_n \\ &= -\, W_j W_n \Delta_n - i\, \varepsilon_{njl} W_l \Delta_n = W_j(-W_n \Delta_n) + \Delta_j \, . \end{aligned} \tag{15}$$

(We have used the algebra and again (13)). Or

$$\underline{W}.\underline{W}\, \Delta_j = W_j \left(W_n \Delta_n \right) . \tag{16}$$

Since the Casimir operator has an inverse and commutes with W_j we may solve for Δ_j

$$\Delta_j = W_j \left(\frac{1}{\underline{W}.\underline{W}}\, W_n \Delta_n \right) \tag{17}$$

i.e. Δ_j is the variation of a local insertion:

$$\Delta_j = W_j\, \hat{\Delta} \tag{18}$$

and is itself local. This fact is sufficient for repairing the WI (8) which was damaged by the subtraction procedure. Eq. (10) reads now

$$W_k \Gamma = W_k \hat{\Delta} + O(\hbar\, \Delta_k) \tag{19}$$

hence

$$W_k(\Gamma - \hat{\Delta}) = O(\hbar\, \Delta_k) \tag{20}$$

and the violation of the symmetry has been pushed to the order $\hbar\Delta_k$ (instead of Δ_k). Since $\hat{\Delta}$ is local and has dimension ≤ 4 it can be absorbed as a counterterm into the interaction with which we calculate diagrams. Recursively we can by this procedure establish the WI to all orders. The remaining free parameters are those of the symmetric counterterms and are to be fixed by normalization conditions. The symmetry requirement and the normalization conditions fixed uniquely the theory.

2. BRS Transformations, Slavnov Identity

As the first example of a non-linear field transformation we discuss the Becchi-Rouet-Stora (BRS) transformations in a pure Yang-Mills theory with simple gauge group. They read [2]

$$s A_\mu = \partial_\mu c_+ + i[c_+, A_\mu], \qquad sB = 0$$
$$s c_+ = i c_+ c_+, \qquad \varphi \equiv \varphi^k \tau^k \text{ for } \varphi = A, B, c_\pm \tag{21}$$
$$s c_- = B,$$

(The τ^k generate the fundamental representation; $c_\pm$ are the anti-commuting Faddeev-Popov fields, B is a Lagrange multiplier field).

We observe first that

$$s^2 \varphi = 0 \qquad \text{for } \varphi = A, B, c_\pm, \tag{22}$$

i.e. the transformations s are nilpotent on the fields. Conversely we convince ourselves that with the ansatz

$$\begin{aligned} s^2 c_+^l &= -\tfrac{1}{2} x^{l[kj]} c_+^k c_+^j , \\ s A_\mu^l &= a^{lk} \partial_\mu c_+^k + i b^{lkm} c_+^k A_\mu^m \end{aligned} \tag{23}$$

the postulate of nilpotency requires $x^{\cdots}$ to be the structure constants of a Lie algebra and $a \simeq 1$, $b \simeq$ adj. by a redefinition of the fields. Hence the nilpotency of the transformations embodies the algebraic structure on the level of the fields. The hope is that on the level of functionals it will also determine the theory.

Since on the gauge field A the BRS-transformations look like local gauge transformations an invariant in terms of A is just the Yang-Mills Lagrangean

$$\begin{aligned} \Gamma_{YM} &= -\frac{1}{4g^2} \operatorname{Tr} \int dx\, F_{\mu\nu} F^{\mu\nu}, \\ F_{\mu\nu} &\equiv \partial_\mu A_\nu - \partial_\nu A_\mu - i[A_\mu, A_\nu] . \end{aligned} \tag{24}$$

In order to obtain a propagator for the field A gauge fixing is required

$$\Gamma_{g.f.} = \operatorname{Tr} \int dx \left(\frac{\alpha}{2} B^2 + B \partial A \right), \tag{25}$$

which can be incorporated in a BRS-invariant fashion:

$$\begin{aligned} \Gamma_{\phi\pi} &= \operatorname{Tr} \int dx\, s\left(\frac{\alpha}{2} c_- B + c_- \partial A \right) \\ &= \operatorname{Tr} \int dx \left(\frac{\alpha}{2} B^2 + B\partial A - c_- \partial(\partial c_+ + i[c_+, A]) \right) . \end{aligned} \tag{26}$$

The main problem in quantizing the BRS-symmetry consists in giving a meaning to those field variations which are composite operators. This can be done by coupling them to external fields and taking care of them as of any other interaction vertex in the theory: they contribute to Feynman rules, power counting and subtractions, admit counterterms, etc.

$$\Gamma_{ext.f.} = \operatorname{Tr} \int dx \left(\rho^\mu s A_\mu + \sigma s c_+ \right) . \tag{27}$$

Considering the external fields as invariant under s we can express the naive BRS-invariance of

$$\Gamma_{cl} = \Gamma_{YM} + \Gamma_{\phi\pi} + \Gamma_{ext.f.} \tag{28}$$

by writing

$$s\Gamma_{cl} = Tr\int(sA\frac{\delta\Gamma}{\delta A} + sc_+\frac{\delta\Gamma}{\delta c_+} + B\frac{\delta\Gamma}{\delta c_-}) = 0 \;. \tag{29}$$

Now, it is suggestive - and can indeed be rigorously justified (cf. b. Sect. 4) - to rewrite this as a Γ_{cl} non-linear equation

$$\mathcal{S}(\Gamma_{cl}) \equiv Tr\int(\frac{\delta\Gamma_{cl}}{\delta\rho}\frac{\delta\Gamma_{cl}}{\delta A} + \frac{\delta\Gamma_{cl}}{\delta\sigma}\frac{\delta\Gamma_{cl}}{\delta c_+} + B\frac{\delta\Gamma_{cl}}{\delta c_-}) = 0 \;, \tag{30}$$

the Slavnov-identity. Our task is to solve (30) when Γ_{cl} has been replaced by

$$\Gamma = \Gamma_{cl} + \hbar\Gamma^{(1)} + \hbar\Gamma^{(2)} + \cdots \;, \tag{7}$$

the generating functional for vertex functions.

To begin with we observe that any functional Γ can be split into

$$\Gamma = \Gamma_{g.f.} + \bar{\Gamma} \tag{31}$$

(where $\bar{\Gamma}$ does not depend on B) by imposing the gauge condition

$$\frac{\delta\Gamma}{\delta B} = \alpha B + \partial A \;. \tag{32}$$

For, this condition is linear in the quantized fields and can then be satisfied naively in many renormalization schemes. Using (31) we rewrite the Slavnov-identiy as

$$\begin{aligned}\mathcal{S}(\Gamma) &\equiv Tr\int(\frac{\delta\bar\Gamma}{\delta\rho}(\frac{\delta\bar\Gamma}{\delta A} - \partial B) + \frac{\delta\bar\Gamma}{\delta\sigma}\frac{\delta\bar\Gamma}{\delta c_+} + B\frac{\delta\bar\Gamma}{\delta c_-}) \\ &= Tr\int(B(\partial\frac{\delta\bar\Gamma}{\delta\rho} + \frac{\delta\bar\Gamma}{\delta c_-}) + \frac{\delta\bar\Gamma}{\delta\rho}\frac{\delta\bar\Gamma}{\delta A} + \frac{\delta\bar\Gamma}{\delta\sigma}\frac{\delta\bar\Gamma}{\delta c_+}) \;.\end{aligned} \tag{33}$$

If we want to achieve $\mathcal{S}(\Gamma)=0$ we have a <u>necessary</u> condition: differentiation with respect to B of (33) yields

$$\mathcal{G}\Gamma \equiv \partial\frac{\delta\bar\Gamma}{\delta\rho} + \frac{\delta\bar\Gamma}{\delta c_-} = 0 \tag{34}$$

i.e., the functional $\bar\Gamma$ depends on c_- and ρ only through the special combination

$$\eta_\mu = \rho_\mu + \partial_\mu c_- \;. \tag{35}$$

(N.B.: (34) is just the equation of motion of the ghost c_-). Again, linear field equations, like (34), can be naively implemented in many renormalization schemes, hence we adopt one of this type and rewrite now the Slavnov-identity as

$$\begin{aligned}\mathcal{S}(\Gamma) &\equiv Tr\int(\frac{\delta\bar\Gamma}{\delta\eta}\frac{\delta\bar\Gamma}{\delta A} + \frac{\delta\bar\Gamma}{\delta\sigma}\frac{\delta\bar\Gamma}{\delta c_+}) = 0 \\ &= \tfrac{1}{2}\, B_{\bar\Gamma}\bar\Gamma \qquad \text{with}\end{aligned} \tag{36}$$

$$\mathcal{B}_{\bar\Gamma} \equiv \mathrm{Tr}\int\left(\frac{\delta\bar\Gamma}{\delta\eta}\frac{\delta}{\delta A} + \frac{\delta\bar\Gamma}{\delta A}\frac{\delta}{\delta\eta} + \frac{\delta\bar\Gamma}{\delta\sigma}\frac{\delta}{\delta c_+} + \frac{\delta\bar\Gamma}{\delta c_+}\frac{\delta}{\delta\sigma}\right) . \tag{37}$$

That this seemingly trivial linearization of the Slavnov-identity is, in fact, of true importance is revealed by the following properties of the $\mathcal{B}$'s

$$\mathcal{B}_\gamma \mathcal{B}_\gamma \gamma = 0 \qquad \forall\gamma , \tag{38}$$

$$\mathcal{B}_\gamma \mathcal{B}_\gamma = 0 \qquad \text{if} \quad \mathcal{B}_\gamma \gamma = 0 . \tag{39}$$

It will be immediately clear that these relations express the nilpotency of the BRS-transformations on the functional level. Indeed, the action principle (App.) tells us

$$\Lambda(\Gamma) = \Delta\cdot\Gamma \tag{40}$$

Using on the l.h.s. (36) and on the r.h.s. (A.3) we find

$$\frac{1}{2}\mathcal{B}_{\bar\Gamma}\bar\Gamma = \Delta + O(\hbar\Delta) \tag{41}$$

Acting with $\mathcal{B}_{\bar\Gamma}$ on this equation and employing (38) there follows

$$0 = \mathcal{B}_{\bar\Gamma}\Delta + O(\hbar\Delta) , \tag{42}$$

hence to lowest order in $\hbar$

$$\mathcal{B}_{\bar\Gamma_{cl}}\Delta = 0 . \tag{43}$$

With the notation $b = \mathcal{B}_{\bar\Gamma_{cl}}$ this consistency condition for the BRS-transformation is written as

$$b\,\Delta = 0 . \tag{44}$$

We note that b is completely known (in particular that the external fields transform under b) and that it is nilpotent

$$b^2 = 0 \tag{45}$$

due to (39) and (29). Solving (44) is not easy and the actual solution represents a mile-stone in the history of Yang-Mills theory [2]. The solution of (44) reads

$$\Delta = b\hat\Delta + r\,\mathrm{Tr}\int \varepsilon_{\mu\nu\rho\sigma}\, c_+ \partial^\mu\left(\partial^\nu A^\rho A^\sigma - \frac{i}{2} A^\nu A^\rho A^\sigma\right) \tag{46}$$

where the second term is the celebrated chiral anomaly. Due to parity we have r = 0 in our model and BRS-invariance can thus be restored: e.g. in one loop we define

$$\hat\Gamma = \Gamma - \hat\Delta \tag{47}$$

we know

$$\frac{\delta\hat{\Delta}}{\delta B} = 0 \;, \qquad G\hat{\Delta} = 0 \; . \tag{48}$$

$$\begin{aligned} \Lambda(\hat{\Gamma}) &= \tfrac{1}{2} B_{\hat{\Gamma}} \hat{\Gamma} = \tfrac{1}{2} B_{\bar{\Gamma}} \bar{\Gamma} + \tfrac{1}{2} B_{-\hat{\Delta}} \bar{\Gamma} + \tfrac{1}{2} B_{\bar{\Gamma}} (-\hat{\Delta}) + O(\hbar^2) \\ &= \Delta - b\hat{\Delta} + O(\hbar^2) \\ &= O(\hbar^2) \; . \end{aligned} \tag{49}$$

Recursively we can thus establish BRS-invariance to all orders.

3. Wess-Zumino Model without Auxiliary Fields

N = 1 supersymmetry transformations are linear and close off-shell (i.e. without use of equations of motion) due to the existence of so-called auxiliary fields. The simplest example is provided by the Wess-Zumino model [3]. It has a classical action

$$\begin{aligned} \Gamma_{inv} = \int \Big(\partial A \partial \bar{A} + \tfrac{i}{4} \psi \overleftrightarrow{\partial\!\!\!/} \bar{\psi} + F\bar{F} &+ m\big(\tfrac{1}{4}\psi\psi - AF + c.c.\big) \\ &+ g\big(\tfrac{1}{2} A\psi\psi - A^2 F + c.c.\big)\Big) \end{aligned} \tag{50}$$

invariant under the supersymmetry transformations

$$\begin{aligned} \delta_\alpha A &= \psi_\alpha & \qquad \delta_\alpha \bar{A} &= 0 \\ \delta_\alpha \psi^\beta &= 2\delta_\alpha{}^\beta F & \qquad \delta_\alpha \bar{\psi}_{\dot{\alpha}} &= 2i\,\partial_{\alpha\dot{\alpha}} \bar{A} \\ \delta_\alpha F &= 0 & \qquad \delta_\alpha \bar{F} &= -i\,\partial_{\alpha\dot{\alpha}} \bar{\psi}^{\dot{\alpha}} \end{aligned} \tag{51}$$

(We use Weyl spinors, cf. [4] for notation and conventions). F is the auxiliary field because it can be algebraically expressed in terms of the other fields (on shell)

$$\frac{\delta\Gamma}{\delta F} = F - m\bar{A} - g\bar{A}^2 \;, \qquad F \overset{*}{=} m\bar{A} + g\bar{A}^2 \; . \tag{52}$$

Inserting this into (50), (51) we obtain

$$\begin{aligned} \Gamma'_{inv} = \int \Big(\partial A \partial \bar{A} &+ \tfrac{i}{4} \psi \overleftrightarrow{\partial\!\!\!/} \bar{\psi} - m^2 A\bar{A} - g^2 (A\bar{A})^2 \\ &+ m\big(\tfrac{1}{4}\psi\psi - g A\bar{A}^2 + c.c.\big) + \tfrac{1}{2} g\big(\psi\psi A + c.c.\big)\Big) \end{aligned} \tag{53}$$

which is naively invariant under

$$\begin{aligned} \delta_\alpha A &= \psi_\alpha & \qquad \delta_\alpha \bar{A} &= 0 \\ \delta_\alpha \psi^\beta &= 2\delta_\alpha{}^\beta \big(m\bar{A} + g\bar{A}^2\big) & \qquad \delta_\alpha \bar{\psi}_{\dot{\alpha}} &= 2i\,\partial_{\alpha\dot{\alpha}} \bar{A} \; . \end{aligned} \tag{54}$$

Γ_{inv} is, of course, a perfectly legitimate action but the transformations (54) are now non-linear and their anti-commutators close only if one uses the equation of motion for the spinor field, i.e. on-shell. Whereas the renormalization of (50), (51) is by now textbook wisdom [4], that of (53), (54) is not quite straightforward.

In analogy to the previous case we introduce [5] an external source u for the non-linear variation

$$\Gamma = \Gamma'_{inv} + \int (2m(gA^2 + mA) + c.c.) \tag{55}$$

and calculate the variation of Γ in the bilinear fashion known from BRS

$$\begin{aligned} \delta_\alpha \Gamma &\equiv \int dx \left(\psi_\alpha \frac{\delta}{\delta A} + \frac{\delta \Gamma}{\delta \bar{m}} \frac{\delta}{\delta \psi^\alpha} - 2i \sigma_{\alpha\dot{\alpha}} \bar{A} \frac{\delta}{\delta \bar{\psi}_{\dot{\alpha}}} \right) \Gamma \\ &= 2 \int dx \; m (2g A \psi_\alpha + m \psi_\alpha) \end{aligned} \tag{56}$$

which is again composite. At this point one could introduce a new source, calculate the variation, find a composite object, etc. - potentially an infinite sequence. But we observe that the spinor equation of motion reads

$$\frac{\delta \Gamma}{\delta \psi} = \frac{i}{2} \sigma \bar{\psi} + \frac{m}{2} \psi + g A \psi \tag{57}$$

which is just what occurs in (56). Eliminating the breaking term by (57) we end up with

$$\begin{aligned} W_\alpha(\Gamma) \equiv -i \int dx \Big(\psi_\alpha \frac{\delta \Gamma}{\delta A} + \left(\frac{\delta \Gamma}{\delta \bar{m}} - 4m \right) \frac{\delta \Gamma}{\delta \psi^\alpha} \\ - 2i \sigma_{\alpha\dot{\alpha}} \bar{A} \frac{\delta \Gamma}{\delta \bar{\psi}_{\dot{\alpha}}} + 2im \sigma_{\alpha\dot{\alpha}} \bar{\psi}^{\dot{\alpha}} \Big) = 0 \end{aligned} \tag{58}$$

Due to the non-linearity of the field transformations we have a Γ-bilinear WI; but due to the conspiracy of (56) and (54) we need only <u>one</u> source and have just a harmless inhomogeneous term in (58); (harmless because it is linear in the quantized field).

If we wish to set up consistency conditions for

$$W_\alpha(\Gamma) = \Delta_\alpha \cdot \Gamma \tag{59}$$

we have to know how an <u>insertion</u> transforms under supersymmetry and what the algebra of those transformations with $W_\alpha(\Gamma)$ is. The linearization of the WI-operator, which was so suggestive in the BRS-case (cf. transition from (36) to (37)), should in fact be viewed as answering the question: how does an insertion transform? In order to find that out we recall that insertions are differential vertex operations [6] and consider an infinitesimal variation of the action

$$\Gamma^\Delta = \Gamma + \varepsilon \Delta , \tag{60}$$

$$\begin{aligned} W_\alpha(\Gamma^\Delta) &= W_\alpha(\Gamma) + \varepsilon W_\alpha^\Gamma \Delta + o(\varepsilon^2) , \\ W_\alpha^\Gamma &\equiv -i \int \left(\psi_\alpha \frac{\delta}{\delta A} + \left(\frac{\delta \Gamma}{\delta \bar{m}} - 4m \right) \frac{\delta}{\delta \psi^\alpha} + \frac{\delta \Gamma}{\delta \psi^\alpha} \frac{\delta}{\delta \bar{m}} - 2i \sigma_{\alpha\dot{\alpha}} \bar{A} \frac{\delta}{\delta \bar{\psi}_{\dot{\alpha}}} \right) . \end{aligned} \tag{61}$$

(Cf. with BRS: W_α^Γ is the analog of $B_{\bar{\Gamma}}$ (37)). One can check by straightforward calculation that the W_α^Γ's and $W_\alpha(\Gamma)$ satisfy the following algebra

$$\begin{aligned} W_\alpha^\Gamma W_\beta(\Gamma) + W_\beta^\Gamma W_\alpha(\Gamma) &= 0 , \\ \bar{W}_{\dot{\alpha}}^\Gamma \bar{W}_{\dot{\beta}}(\Gamma) + \bar{W}_{\dot{\beta}}^\Gamma \bar{W}_{\dot{\alpha}}(\Gamma) &= 0 , \\ W_\alpha^\Gamma \bar{W}_{\dot{\beta}}(\Gamma) + \bar{W}_{\dot{\beta}}^\Gamma W_\alpha(\Gamma) &= 2 \sigma^\mu_{\alpha\dot{\beta}} W_\mu \Gamma . \end{aligned} \tag{62}$$

Here W_μ denotes the WI-operator for the translations (which yields zero if Γ is the action or the vertex functional because those are translation invariant).

The higher order machinery is now prepared: we use (A.3) for (59), act suitably with W_α^Γ resp. $\bar{W}_{\dot\alpha}^\Gamma$ on (59), and use (62). We obtain for the insertions $\Delta_\alpha, \bar\Delta_{\dot\alpha}$ the following consistency conditions

$$\begin{aligned} \mathcal{W}_\alpha \Delta_\beta + \mathcal{W}_\beta \Delta_\alpha &= 0 \\ \bar{\mathcal{W}}_{\dot\alpha} \bar\Delta_{\dot\beta} + \bar{\mathcal{W}}_{\dot\beta} \bar\Delta_{\dot\alpha} &= 0 \\ \mathcal{W}_\alpha \bar\Delta_{\dot\beta} + \bar{\mathcal{W}}_{\dot\beta} \Delta_\alpha &= 0 \end{aligned} \tag{63}$$

($\mathcal{W}_\alpha \equiv W_\alpha^{\Gamma_{cl}}$ and we calculated to lowest order in $\hbar$). It is now a matter of some algebraic work [5] to show that as a consequence of (63) one has

$$\begin{aligned} \Delta_\alpha &= \mathcal{W}_\alpha \hat\Delta \\ \bar\Delta_{\dot\alpha} &= \bar{\mathcal{W}}_{\dot\alpha} \hat\Delta \end{aligned} \tag{64}$$

i.e. the cohomology of the supersymmetry defined by (58) is still trivial (as it was in the linear realization [7]). As usual we can absorb $\hat\Delta$ into Γ and proceed by induction to show that the WI (58) holds to all orders; i.e. supersymmetry in the version of (58) has again defined the model and by studying e.g. the Callan-Symanzik equation one comes to the conclusion that physicswise the linear version [3] and the non-linear version (58) are equivalent[5] .

4. General Formalism for Non-linear Symmetry Transformations [8]

Up to now we based our discussion on Γ . The reason for this is twofold:

(1) the lowest term of Γ in its $\hbar$-expansion can be identified with the classical action on which all manipulations are most familiar;

(2) Γ governs the renormalization: connected or general Green's functions are divergent only if their one-particle-irreducible parts are so. But the action principle, the equations of motion, etc., are generally proved with the help of Z (or Z_c) - the generating functional for (connected) Green's functions. In fact, non-linear field transformations are easiest formulated on Z (or Z_c) and not on Γ.

So let us study, again first in the linear case, the transition from Γ to Z_c and to Z. Let

$$\delta_k \phi_a = i (t_k)_{ab} \phi_b \tag{65}$$

be the linear transformation law for the elementary fields ϕ_a.

$$W_k \Gamma \equiv -i \int i\, \phi_a (t_k^T)_{ab} \frac{\delta \Gamma}{\delta \phi_b} \tag{66}$$

is the WI-operator. Z_c is defined via the Legendre transformation

$$Z_c(j) = \Gamma(\phi) + \int j_a \phi_a \tag{67}$$

where

$$j_a = -\frac{\delta\Gamma}{\delta\phi_a} \quad . \tag{68}$$

The inverse transformation is given by

$$\frac{\delta Z_c}{\delta j_a} = \phi_a \quad . \tag{69}$$

On Z_c the WI-operator reads

$$W_k Z_c = -i\int i \frac{\delta Z_c}{\delta j_a} (t_k^T)_{ab}(-j_b) = -\int j_a (t_k)_{ab} \frac{\delta Z_c}{\delta j_b} \quad . \tag{70}$$

On
$$Z = e^{iZ_c} \tag{71}$$

it has the same form. Before proceeding let us note that for composite operators ϕ_c with source j_c the Legendre transformation is given by

$$\frac{\delta\Gamma}{\delta j_c} = \frac{\delta Z_c}{\delta j_c} \tag{72}$$

in contrast to (68), (69) since the Legendre transformation is performed only with respect to elementary fields, i.e. those operators producing one-particle poles, with respect to which one can speak of one-particle reducibility or irreducibility. Eq.(72) will soon be seen to cause the different form a WI-operator has on Z_c and on Γ.

Let now ϕ denote <u>all</u> elementary fields of the theory [8]. Suppose the polynomials $P_k(\phi)$ occur as non-linear variations

$$\delta_k \phi = P_k(\phi) \quad . \tag{73}$$

Then we introduce sources j_c for the P_k, need possibly sources for the double variations, etc. Hence we collect all fields and sources in

$$\Phi = \begin{pmatrix} \phi \\ \phi_c \end{pmatrix} \quad , \quad J = \begin{pmatrix} j \\ j_c \end{pmatrix} \quad , \tag{74}$$

assume that the field transformation law is linear in Φ

$$\delta_k \Phi = i t_k \Phi =: i \begin{pmatrix} t_{\phi\phi} & t_{\phi c} \\ t_{c\phi} & t_{cc} \end{pmatrix} \begin{pmatrix} \phi \\ \phi_c \end{pmatrix} \tag{75}$$

and admits a certain algebra

$$[\delta_k, \delta_l] = i f_{klm} \delta_m \quad . \tag{76}$$

We formulate the WI-operator for these transformations on Z_c in analogy to (70) as

$$W_k Z_c \equiv -\int J_a (t_k)_{ab} \frac{\delta Z_c}{\delta J_b} \quad . \tag{77}$$

The Legendre transformation with respect to the elementary fields

$$\frac{\delta Z_c}{\delta j_a} = \phi_a = \phi_a(j, j_c) \qquad \Gamma(\phi, j_c) = Z_c - \int j_a \phi_a \tag{78}$$

inverse: $\quad \dfrac{\delta \Gamma}{\delta \phi_a} = -j_a(\phi, j_c)$

leads to the WI-operator

$$\begin{aligned} W_k(\Gamma) &\equiv i\int i\left(-\frac{\delta\Gamma}{\delta\phi} \quad j_c\right)\begin{pmatrix} t_{\phi\phi} & t_{\phi c} \\ t_{c\phi} & t_{cc}\end{pmatrix}_k \begin{pmatrix}\phi \\ \frac{\delta\Gamma}{\delta j_c}\end{pmatrix} \\ &= -i\int i\left(\phi \quad \frac{\delta\Gamma}{\delta j_c}\right)\begin{pmatrix} t_{\phi\phi} & t_{\phi c} \\ t_{c\phi} & t_{cc}\end{pmatrix}_k^T \begin{pmatrix}\frac{\delta\Gamma}{\delta\phi} \\ -j_c\end{pmatrix} \end{aligned} \tag{79}$$

which is non-linear in Γ. For consistency conditions we have to know how an insertion transforms (cf. a. Sect. 3)

$$\Gamma \to \Gamma + \varepsilon\Delta \, , \tag{80}$$

$$W_k^\Gamma \Delta = -i\int i\left(\phi \quad \frac{\delta\Gamma}{\delta j_c}\right) t_k^T \begin{pmatrix}\frac{\delta\Delta}{\delta\phi} \\ 0\end{pmatrix} + i\left(0 \quad \frac{\delta\Delta}{\delta j_c}\right) t_k^T \begin{pmatrix}\frac{\delta\Gamma}{\delta\phi} \\ -j_c\end{pmatrix} \quad . \tag{81}$$

The algebra of the transformations (76) leads to the algebra of the WI-operators

$$W_k^\Gamma W_l(\Gamma) - W_l^\Gamma W_k(\Gamma) = i f_{klj} W_j(\Gamma) \quad . \tag{82}$$

From the examples above it should by now be clear that this algebra leads to the consistency conditions

$$\mathcal{W}_k \Delta_l - \mathcal{W}_l \Delta_k = i f_{klj} \Delta_j \tag{83}$$

for the insertions appearing in

$$W_k(\Gamma) = \Delta_k \cdot \Gamma \tag{84}$$

$(\mathcal{W}_k \equiv W_k^{\Gamma_{cl}})$. It should also be clear that the desired WI

$$W_k(\Gamma) = 0 \tag{85}$$

can be proved if and only if the solution of (83) is the trivial one

$$\Delta_k = \mathcal{W}_k \hat{\Delta} \tag{86}$$

This finishes the presentation of the general case of non-linear symmetry transformations.

APPENDIX: Notations, the Action Principle

Γ is the generating functional for vertex functions,

Z_c is the generating functional for connected Green's functions,

Z is the generating functional for general Green's functions.

$$\Gamma = \Gamma(\varphi, j_c; m, g) \tag{A.1}$$

φ	elementary fields
j_c	sources for composite operators
m,g	parameters

The action principle [9] states

$$\nabla \mathcal{F} = \Delta \cdot \mathcal{F} \qquad \mathcal{F} = \Gamma, Z_c, Z \tag{A.2}$$

i.e. operating with a differential operator on Γ, Z_c or Z one obtains a local insertion of specified power counting. If insertions are defined as Zimmermann-Lowenstein normal products [10] one has the following examples

∇	∂_g	$m\partial_m$	$\frac{\delta}{\delta\varphi}$	$\int\varphi'\frac{\delta}{\delta\varphi}$	$\frac{\delta}{\delta j_c}$
Δ^{ρ}_{δ}	4	4	$4-\rho(\varphi)$	$4-\rho(\varphi)+\rho(\varphi')$	$4-\rho(j_c)$
	4	4	$4-d(\varphi)$	$4-d(\varphi)+d(\varphi')$	$4-d(j_c)$

Another simple but very powerful relation is the following

$$\Delta \cdot \Gamma = \Delta + O(\hbar\Delta) \tag{A.3}$$

Here $\Delta\cdot\Gamma$ denotes the generating functional for all vertex functions housing the special vertex Δ. The Eq.(A.3) means that in the loop expansion the first term is the trivial one: the vertex Δ itself. Example:

$$\Delta = \int dx\, \varphi^2$$

+ genuine loop diagrams

i.e. only the 2-point function $(\Delta\cdot\Gamma)_{12}$ has a tree contribution, all others start with genuine loops.

REFERENCES

[1] G. Velo, A.S. Wightman (eds.), Renormalization Theory, D. Reidel Publ. Co. Dordrecht Holland 1976

[2] C. Becchi, A. Rouet, R. Stora, Ann. of Phys. 98 (1976) 287

[3] J. Wess, B. Zumino, Phys. Lett. 49B (1974) 52

[4] O. Piguet, K. Sibold, Renormalized Supersymmetry, Birkhäuser Boston 1986

[5] O. Piguet, K. Sibold, Nucl. Phys. B253 (1985) 269

[6] J.H. Lowenstein, Comm. Math. Phys. 24 (1971) 1

[7] O. Piguet, M. Schweda, K. Sibold, Nucl. Phys. B174 (1980) 183

[8] P. Breitenlohner, D. Maison in Supersymmetry and its Applications: Superstrings, Anomalies and Supergravity (eds. G.W. Gibbons, S.W. Hawking, P.K. Townsend) Cambridge 1986

[9] Y.M.P. Lam, Phys. Rev. D6 (1972) 2145, 2161
T.E. Clark, J.H. Lowenstein, Nucl. Phys. B113 (1976) 109

[10] J.H. Lowenstein, W. Zimmermann, Comm. Math. Phys. 44 (1975) 73
J.H. Lowenstein, Comm. Math. Phys. 47 (1976) 53

SUPERSPACE RENORMALIZATION OF N = 1, d = 4 SUPERSYMMETRIC GAUGE THEORIES

Olivier Piguet
Theory Division, CERN, CH-1211 Geneva 23, Switzerland

Contents:

These two somewhat unrelated talks deal with the renormalization of N = 1 supersymmetric gauge theories in four-dimensional space-time. We are working in the superfield formalism, i.e., in a linear realization of supersymmetry, with a supersymmetry- invariant gauge-fixing condition. This is to be contrasted with the Wess-Zumino gauge approach, where the non-linear realization of supersymmetry causes some difficulties which are still awaiting a complete solution[1)]. However, although renormalization is made simpler by the superfield approach[2),3)], a substantial price has to be paid, due to the fact that the gauge superfield is dimensionless and massless.

Indeed, the consequences of this fact are, first, the occurrence of a non-linear renormalization of the gauge superfield[3),4)], a phenomenon which was also met later on in the study of two-dimensional σ-models[5)-7)], and, second, off-shell infra- red singularities due to a propagator of the form $1/k^4$ for this same gauge superfield. This is the subject of the first talk, where we show that the infinite set of arbitrary parameters describing the non-linear field renormalization[*)] are gauge parameters, and thus do not contribute to physical quantities like Green functions of gauge-invariant operators. The method of the proof consists of allowing these parameters to transform under BRS and of proving the corresponding Slavnov identity. This procedure is explained in Ref. 9) for the case of gauge parameters in ordinary Yang-Mills theories, and was in fact already advocated in Ref. 10). The application to the supersymmetric case we discuss here was given in Refs. 3) and 4). We also briefly describe in this first talk the use of this procedure for curing the infra-red singularity, by introducing an infra-red cut-off mass and showing that it is a gauge parameter[11)].

The second talk deals with the problem of finite theories. More precisely, we consider theories with vanishing Callan-Symanzik β-functions, namely conformal invariant theories, which can be interpreted as finite "on the mass-shell". For these, in particular, the Green functions of gauge-invariant operators without

---

*) This phenomenon was also discovered, independently, by explicit one-loop graph computations[8)].

anomalous dimensions, e.g., conserved flavour currents, have no ultra-violet divergences. We shall show that N = 1 super-Yang-Mills theories coupled with matter indeed have vanishing β functions, if they satisfy three conditions which can be checked by simple one-loop computations[12),13)]. These criteria may be expressed in the following way.

(1) The gauge coupling β-function vanishes in the one-loop approximation.

(2) The anomalies of the axial currents associated with the set of chiral invariances of the superpotential, i.e., of the action describing the self-interaction of the matter fields, vanish.

(3) The coupling constants are completely reduced[14)]. In other words, all matter self-interaction coupling constants λ_I can be chosen in a consistent way as functions of the gauge coupling constant g[*)], so that the theory depends only on one coupling constant.

These criteria will be shown to be sufficient for the vanishing of the β-functions. On the other hand, condition (1) is clearly necessary. Condition (3) is also necessary in view of the lower-order calculations of Ref. 15). Let us mention that in the latter reference, as well as in the remaining literature, the vanishing of the anomalous dimensions of the matter fields is required. This is indeed sufficient for the matter self-interaction β-functions to vanish, but in general not necessary. Our three criteria can be seen to be fulfilled[12)] by the extended N = 4 super-Yang-Mills theory, as well as by a class of N = 2 theories, all written in terms of N = 1 superfields. This confirms the known results[16),17)]. The criteria are also satisfied by N = 1 theories with complex representations for the matter fields[13)].

I. - NON-LINEAR FIELD RENORMALIZATION[3),4)]

I.1 Classical Theory

The field content of the theory is given by a set of real gauge superfields $\phi^i(x,\theta,\bar\theta)$ (dimension 0), a set of Lagrange multiplier chiral superfields $B^i(x,\theta)$ (dimension 1) and a set of anticommuting ghost and antighost chiral superfields $c_+^i(x,\theta)$ and $c_-^i(x,\theta)$ (dimensions 0 and 1). No coupling with matter fields will be

---

*) We consider a simple gauge group; thus there is only one gauge coupling constant.

considered in this section. The superscript i is the Yang-Mills index, and we shall use the matrix notation

$$\phi = \phi^i \tau_i, \quad B = B^i \tau_i, \quad c_\pm = c_\pm^i \tau_i \tag{1.1}$$

where the matrices τ_i are the generators of the gauge group in the fundamental representation. Notations and conventions are those of Ref. 3). The gauge group is chosen to be simple.

The BRS transformations may be written as

$$\begin{aligned} se^\phi &= e^\phi c_+ - \bar{c}_+ e^\phi \\ s\phi &= c_+ - \bar{c}_+ + \tfrac{1}{2}[\phi, c_+ + \bar{c}_+] + \dots \equiv Q_s(\phi, c_+) \\ sc_+ &= -c_+^2 \\ sc_- &= B, \quad sB = 0 \end{aligned} \tag{1.2}$$

and are nilpotent:

$$s^2 = 0. \tag{1.3}$$

Introducing external superfields ρ^i and σ^i coupled to the BRS variations of ϕ^i and c_+^i respectively, we can write an action invariant under (1.2) as: [matrix notation (1.1) is used]:

$$\begin{aligned} \Gamma_s(\phi, c_\pm, B, \rho, \sigma) &= -\frac{1}{128g^2} \operatorname{Tr} \int dS F^\alpha F_\alpha \\ &+ \operatorname{Tr} \int dV \{ -\frac{1}{8} (DDc_- + \bar{D}\bar{D}\bar{c}_-) Q_s(\phi, c_+) + \rho Q_s(\phi, c_+) \} \\ &- \operatorname{Tr} \int dS \sigma c_+^2 - \operatorname{Tr} \int d\bar{S} \bar{\sigma} \bar{c}_+^2 \\ &+ \operatorname{Tr} \int dV \{ (DDB + \bar{D}\bar{D}\bar{B}) \phi + \alpha B\bar{B} \} \end{aligned} \tag{1.4}$$

where

$$F_\alpha = \bar{D}\bar{D}e_\alpha, \quad e_\alpha = e^{-\phi} D_\alpha e^{\phi} \tag{1.5}$$

$$dV = d^4x DD\,\bar{D}\bar{D}, \quad dS = d^4x DD, \quad d\bar{S} = d^4x \bar{D}\bar{D} \tag{1.6}$$

D_α, $\bar{D}_{\dot\alpha}$ are the superspace covariant derivatives.

The BRS invariance of (1.4) may be expressed by the Slavnov identity[18)]

$$\begin{aligned} S(\Gamma) =: \; & \mathrm{Tr} \int dV \frac{\delta\Gamma}{\delta\rho}\frac{\delta\Gamma}{\delta\phi} \\ & + \mathrm{Tr} \int dS \left[\frac{\delta\Gamma}{\delta\sigma}\frac{\delta\Gamma}{\delta c_+} + B\frac{\delta\Gamma}{\delta c_-}\right] - \text{c.c.} = 0 \end{aligned} \tag{1.7}$$

and the gauge fixing [last terms in (1.4)] by the (linear) gauge-fixing condition

$$\frac{\delta\Gamma}{\delta B} = \frac{1}{8}\bar{D}\bar{D}\,DD\phi + \alpha\,\bar{D}\bar{D}\,\bar{B}. \tag{1.8}$$

The theory is further specified by supersymmetry and rigid invariance

$$\begin{aligned} & \delta_\alpha\psi = \left(\frac{\partial}{\partial\theta\alpha} + i\sigma^\mu_{\alpha\dot\alpha}\,\bar\theta^{\dot\alpha}\partial_\mu\right)\psi \\ & \delta_{rig}\psi = i\,[\psi,\omega], \quad \omega = \omega^i\tau_i, \quad \omega^i = \text{const.} \\ & \psi = \phi,\ c_\pm,\ B,\ \rho,\ \sigma \end{aligned} \tag{1.9}$$

expressed through the Ward identities[18)]

$$\begin{aligned} & W_\alpha\Gamma = -i\sum_\psi \int \delta_\alpha\psi\,\frac{\delta\Gamma}{\delta\psi} = 0 \\ & W_{rig}\Gamma = -i\sum_\psi \int \delta_{rig}\psi\,\frac{\delta\Gamma}{\delta\psi} = 0 \end{aligned} \tag{1.10}$$

From now on, we <u>define</u> the theory through the functional identities (1.7), (1.8) and (1.10). This is the appropriate way for the extension to the quantized theory.

Whereas the requirements (1.8) and (1.10) are straightforward, we shall see that the action (1.4) is not the most general classical solution of the Slavnov identity (1.7). In order to investigate this, let us consider the following stability problem: the special solution Γ_s (1.4) being given, find the most general form of the perturbed action (ε small)

$$\Gamma = \Gamma_s + \varepsilon\Delta(\phi,\ c_+,\ c_-,\ B,\ \rho,\ \sigma) \tag{1.11}$$

fulfilling all of our requirements, and having its dimension bounded by four in order to preserve power-counting renormalizability. From supersymmetry and rigid invariance (1.10), we know that Δ is a linear combination of superspace integrals of rigid-invariant superfield monomials - there is an infinite set of them, since ϕ is dimensionless.

The gauge-fixing condition (1.8) implies

$$\begin{aligned}\Gamma(\phi,\ c_+,\ c_-,\ B,\ \rho,\ \sigma) &= \bar{\Gamma}(\phi,\ c_+,\ \eta,\ \sigma)\\ &+ \int\{\tfrac{1}{8}(DDB + \bar{D}\bar{D}\bar{B})\phi + \alpha\, B\bar{B}\}\end{aligned} \tag{1.12}$$

with

$$\eta = \rho - \tfrac{1}{8}(DDc_- + \bar{D}\bar{D}\bar{c}_-). \tag{1.13}$$

That the dependence on c_- and ρ occurs through the combination η (1.13) is a consequence of the ghost equation

$$G\Gamma = : [\ \frac{\delta}{\delta c_-} + \frac{1}{8}\ \bar{D}\bar{D}\ DD\,\frac{\delta}{\delta\rho}\ ]\ \Gamma = 0 \tag{1.14}$$

which in turn follows from the gauge-fixing condition and the Slavnov identity.

The ansatz (1.12) allows us to write the Slavnov identity (1.7) in the form

$$S(\Gamma) = \tfrac{1}{2}\ B_{\bar{\Gamma}}\bar{\Gamma} = 0 \tag{1.15}$$

with

$$B_{\bar{\Gamma}} = :\int\{\ \frac{\delta\bar{\Gamma}}{\delta\eta}\frac{\delta}{\delta\phi} + \frac{\delta\bar{\Gamma}}{\delta\phi}\frac{\delta}{\delta\eta} + \frac{\delta\bar{\Gamma}}{\delta\sigma}\frac{\delta}{\delta c_+} + \frac{\delta\bar{\Gamma}}{\delta c_+}\frac{\delta}{\delta\sigma}\} \tag{1.16}$$

(We drop all trace and integration measure symbols.) The functional-dependent linear operator (1.16) obeys the identities

$$B_\gamma\, B_\gamma\, \gamma = 0,\ \forall\, \gamma \tag{1.17}$$

$$B_\gamma^2 = 0, \text{ if } B_\gamma \gamma = 0. \tag{1.18}$$

Γ and Γ_s being each decomposed according to (1.12), (1.11) reads

$$\bar{\Gamma} = \bar{\Gamma}_s + \varepsilon\, \Delta(\phi, c_+, \eta, \sigma). \tag{1.19}$$

Substituting this in the Slavnov identity (1.15) and retaining the terms of first order in ε, we obtain for Δ the equation

$$b\Delta = 0 \tag{1.20}$$

$$\text{with } b =: B_{\bar{\Gamma}_s},\ b^2 = 0 \tag{1.21}$$

[The nilpotency of b follows from (1.18) since Γ_s is a solution of the Slavnov identity.] Note that b, when acting on ϕ and c_+, coincides with the BRS operator s (1.2). But it acts non-trivially on the external fields:

$$b\eta = \frac{\delta \bar{\Gamma} s}{\delta \phi},\ b\sigma = \frac{\delta \bar{\Gamma} s}{\delta c_+} \tag{1.22}$$

To solve (1.20) is a cohomology problem, with the coboundary operator given by (1.21). The most general solution Δ having ghost number 0*) and dimension 4 has the form[4)]

$$\Delta = z\, \Gamma_{SYM}(\phi) + b\, \hat{\Delta}(\phi, c_+, \eta, \sigma) \tag{1.23}$$

where

$$\Gamma_{SYM} = -\frac{1}{128} \mathrm{Tr} \int dS F^\alpha F_\alpha \tag{1.24}$$

is the super-Yang-Mills gauge-invariant action occurring in (1.4), and $\hat{\Delta}$ is an arbitrary local functional of dimension 4 and ghost number -1:

$$\hat{\Delta} = \mathrm{Tr} \int dV f(\phi)\eta - [x\, \mathrm{Tr} \int dS c_+ \sigma + \text{c.c.}] \tag{1.25}$$

with

$$f(\phi) = \sum_{k=1}^{\infty} x_k (\phi)^k \tag{1.26}$$

---

*) The ghost numbers of ϕ, c_+, c_-, ρ, σ are 0, 1, -1, -1, -2 respectively.

or, more precisely:

$$f_i(\phi) = \sum_{k=1}^{\infty} \sum_{\omega=1}^{\Omega_k} x_{k,\omega} \, t^{\omega}_{i(i_1 \ldots i_k)} \, \phi^{i_1} \ldots \phi^{i_k} \tag{1.27}$$

where $t^{\omega}_{i(i_1 \ldots i_k)}$ are the Ω_k invariant tensors of rank k+1, symmetric in their k last indices (rigid invariance is taken into account). z, x and $x_{k,\omega}$ are arbitrary parameters.

Computing $b\hat{\Delta}$ according to the definitions (1.21) and (1.16) we find (integration measures and trace symbols omitted)

$$b\Delta = \int\Big\{f_i \frac{\delta\bar{\Gamma}s}{\delta\phi_i} - \eta_i \partial_j f_i \frac{\delta\bar{\Gamma}s}{\delta\eta_j} + x\Big(c_+ \frac{\delta\bar{\Gamma}s}{\delta c_+} - \sigma \frac{\delta\bar{\Gamma}s}{\delta\sigma}\Big)\Big\} \tag{1.28}$$

with $\partial_j f_i = (\partial/\partial\phi^j) f_i(\phi)$.

Substituting (1.23) into (1.19) yields, at the first order in ε,

$$\bar{\Gamma}(\phi, c_+, \eta, \sigma) = \bar{\Gamma}_s(\hat{\phi}, \hat{c}_+, \hat{\eta}, \hat{\sigma})\big|_{g^2 \to g^2 - \varepsilon z} \tag{1.29}$$

with

$$\begin{aligned} &\hat{\phi}_i = \phi_i + \varepsilon f_i(\phi), \; \hat{\eta}_i = \eta_i - \varepsilon\eta_j \partial_i f_j(\phi) \\ &\hat{c}_+ = (1+\varepsilon x)c_+, \; \hat{\sigma} = (1-\varepsilon x)\sigma. \end{aligned} \tag{1.30}$$

This means that the general solution in the neighbourhood of the special solution $\bar{\Gamma}_s$ is obtained by a coupling constant renormalization $g^2 \to g^2 - \varepsilon z$ and the field substitutions (1.30). For c_+ this is just a usual field amplitude renormalization, but for ϕ we have a generalized, <u>non-linear</u> field amplitude renormalization.

We notice that the sources η and σ for the BRS transformations of ϕ and c_+ are redefined, too: this amounts to a redefinition of the BRS transformation laws (1.2) $s \to \hat{s}$, such that

$$\hat{s}\, e^{\hat{\phi}} = e^{\hat{\phi}} \hat{c}_+ - \hat{c}_+ e^{\hat{\phi}}, \quad \hat{s}\, \hat{c}_+ = -\hat{c}^2_+. \tag{1.31}$$

This is in fact[3)] the most general change of s keeping its nilpotency - which is implicitly contained in the definition of the Slavnov operator [see (1.17) and (1.18)].

The relevance of studying the general solution in the infinitesimal form (1.11) lies in the fact that it yields the general structure of the counterterms of the quantized theory in the perturbative framework.The occurrence of the non-linear renormalization (1.30) was indeed confirmed by explicit one-loop computations[8] which showed the presence of infinities, absorbable only through a non-linear redefinition of ϕ.

One has, however, to look for the general classical solution of the Slavnov identity in finite form, since this is the starting point for the perturbative construction of the quantum theory. It turns out[4] that this general solution is again obtained from a special solution $\bar{\Gamma}_s$ by a substitution exactly as in (1.29), but now with the finite field redefinitions

$$\hat{\phi}_i = F_i(\phi), \quad \hat{\eta}_i = \eta_j \left[\frac{\partial}{\partial\hat{\phi}_i} F_j^{-1}(\hat{\phi})\right]_{\hat{\phi}=F(\phi)}$$
$$\hat{c}_+ = Z_{c_+}\, c_+, \quad \hat{\sigma} = Z_{c_+}^{-1}\, \sigma \qquad (1.32)$$

where [in the short-hand notation (1.26) instead of (1.27)]

$$F(\phi) = Z_\phi\, \phi + \sum_{k\geq 2} a_k(\phi)^k \qquad (1.33)$$

is an arbitrary invertible, dimension 0, function of ϕ. Z_{c_+}, Z_ϕ a_k are arbitrary constants.

We may conclude that the theory depends on infinitely numerous parameters and hence is non-renormalizable! The following formal argument suggests that the parameters a_k are in fact gauge parameters, hence non-physical. (A rigorous proof will be given in the next subsection for the quantized theory.) We first observe that the substitutions (1.32) for ϕ and η (we now take $\hat{c}_+ = c_+$, $\hat{\sigma} = \sigma$) take place in $\bar{\Gamma}$ as defined by (1.12), and not in the whole action Γ_s. But if, after this, we perform the inverse transformation $\phi \to F^{-1}(\phi)$, and similarly for η, in the whole action Γ (this defines a canonical transformation) we arrive at the equivalent action

$$\Gamma'(\phi,c_+,c_-,B,\rho,\sigma) = \bar{\Gamma}_s(\phi,c_+,\eta,\sigma) + \int\left\{\frac{1}{8}\,(DDB + \overline{DD}\,\bar{B})\, F^{-1}(\phi) + \alpha B\bar{B}\right\} \qquad (1.34)$$

We see in this new formulation that the numbers a_k now parametrize the non-linear gauge fixing condition

$$\frac{\delta\Gamma'}{\delta B} = \frac{1}{8}\, \overline{DD}\, DD\, F^{-1}(\phi) + \alpha\, \overline{DD}\, \bar{B} \qquad (1.35)$$

which replaces the linear condition (1.8): they are indeed gauge parameters.

I.2 Renormalization

The quantum theory is described by the generating functional $Z(J_\phi, J_{c_+}, J_{c_-}, J_B, \rho, \sigma)$ of the Green functions, or by

$$Z_c = \frac{\hbar}{i} \log Z \tag{1.36}$$

which generates the connected Green functions, or by the vertex functional

$$\Gamma(\phi, c_+, c_-, B, \rho, \sigma) = Z_c(J_\phi, J_{c_+}, J_{c_-}, J_B, \rho, \sigma) - \int \{J_\phi \phi + J_{c_+} c_+ + J_{c_-} c_- + J_B B\} \tag{1.37}$$

which generates the one-particle irreducible graphs and coincides with the classical action at $\hbar = 0$, $\hbar$ being taken as the perturbation expansion parameter (loop expansion). The theory is defined by requiring the supersymmetry and rigid invariance Ward identities (1.10), the linear gauge fixing condition (1.8) and the Slavnov identity (1.7). The latter reads, for the Green functional Z, in short-hand notation*):

$$SZ =: \int \{-J_\phi \frac{\delta}{\delta\rho} + J_{c_+} \frac{\delta}{\delta\sigma} + J_{c_-} \frac{\delta}{\delta J_B}\} Z \sim 0 \tag{1.38}$$

with $S^2 = 0$.

As we have seen, the theory depends on the infinite set of parameters a_k describing the general non-linear ϕ-field renormalization. We will now show that this dependence has the peculiar form

$$\frac{\partial}{\partial a_k} Z \sim S(\Delta_k . Z) \tag{1.39}$$

where Δ_k are some insertions with ghost number -1, or, equivalently for Γ:

$$\frac{\partial}{\partial a_k} \Gamma \sim B_{\bar{\Gamma}} (\Delta_k . \Gamma) \tag{1.40}$$

with $B_{\bar{\Gamma}}$ defined by (1.16). The consequence of (1.39) is that the S-matrix - if it can be defined - is independent of the a_k's. More generally one can define the Green

---

*) In order to avoid here any infra-red problem caused by the presence of dimensionless fields we introduce masses which preserve supersymmetry but break BRS invariance. Hence the Slavnov identity can only hold up to soft terms: this is expressed by the sign $\sim$ in all subsequent identities.

functions of gauge invariant operators Q_a through the introduction of BRS invariant external fields q_a. Their generating functional

$$Z_{inv}(q) = Z(J,\rho,\sigma,q)\Big|_{J=\rho=\sigma=0} \tag{1.41}$$

is then a_k-independent:

$$\frac{\partial}{\partial a_k} Z_{inv}(q) = [S(\Delta_k.Z)]_{J=\rho=\sigma=0} = 0. \tag{1.42}$$

The validity of (1.39) or (1.40) is easily checked in the classical approximation. [In the infinitesimal form (1.11) this directly follows from the fact that the a_k- (or x_k-) dependent part of the perturbation Δ is of the form $B_{\bar{\Gamma}}\int(\phi)^k\eta$ as can be seen from Eqs. (1.23)-(1.26).]

In order to extend the property (1.40) to the quantum theory, we require the latter to obey the new Slavnov identity

$$S(\Gamma) = S_{old}(\Gamma) + \sum_k x_k \frac{\partial\Gamma}{\partial a_k} \sim 0 \tag{1.43}$$

where we have introduced an infinite set of anticommuting parameters x_k. This amounts to considering the a_k's as transforming under BRS (in a way respecting the nilpotency):

$$s\, a_k = x_k, \quad sx_k = 0 \tag{1.44}$$

It is checked that this works by differentiating (1.43) with respect to x_k, which yields

$$\frac{\partial\Gamma}{\partial a_k} - B_{\bar{\Gamma}} \frac{\partial\Gamma}{\partial x_k} \sim 0 \tag{1.45}$$

with

$$B_{\bar{\Gamma}} = B_{\bar{\Gamma}}^{old} + \sum_k x_k \frac{\partial}{\partial a_k} . \tag{1.46}$$

Equation (1.45) indeed reproduces (1.40) with the identification

$$\Delta_k.\Gamma = \frac{\partial\Gamma}{\partial x_k} . \tag{1.47}$$

The construction of an x_k-dependent classical action fulfilling the new Slavnov identity is straightforward. By standard arguments[18] we know that the construction is then feasible at all orders of perturbation theory if the cohomology equation

$$b\Delta = 0 \tag{1.48}$$

admits only trivial solutions

$$\Delta = b\hat{\Delta}. \tag{1.49}$$

Here, Δ being a local functional $\Delta(\phi, c_+, \eta, \sigma, a_k, x_k)$ of dimension four and ghost number one, the solution $\hat{\Delta}$ must be local, too, with dimension four and ghost number zero. The coboundary operator is

$$b = B_{\bar{\Gamma}_{\text{classical}}}, \quad b^2 = 0. \tag{1.50}$$

In particular

$$ba_k = x_k, \; bx_k = 0. \tag{1.51}$$

Concerning the x_k and a_k dependence of Δ, the cohomology is that of polynomials*), and is thus trivial:

$$\Delta = b\hat{\Delta}_1 + \Delta_2 \tag{1.52}$$

with

$$b\Delta_2 = 0, \; \frac{\partial \Delta_2}{\partial a_k} = \frac{\partial \Delta_2}{\partial x_k} = 0. \tag{1.53}$$

The remaining cohomology problem (1.53) is well known[3),19)]: the only non-trivial solution is the chiral anomaly, which we assume to be absent.

We have proved in this way the possibility of constructing a theory obeying the new Slavnov identity (1.43). In other words there always exists a set of insertions Δ_k such that the physical a_k independence condition (1.39) holds.

---

*) The a_k's play the role of coupling constants. Since they couple with increasing powers of ϕ as k increases, any term of a given order in $\hbar$ (i.e., given number of loops) and of a given degree in ϕ can only depend on a finite number of a_k.

I.3 The off-shell infra-red problem[3),11)]

Since the $\theta = 0$ component of the gauge superfield ϕ is of dimension zero, and massless in the case of strict gauge invariance, it has a propagator of the form $1/k^4$, which causes infra-red (IR) divergent Green functions.

In order to cure this disease we take advantage of the freedom of doing an arbitrary field redefinition (1.32). We choose a θ-dependent redefinition

$$\phi \to F(\phi) = (1 + \tfrac{1}{2}\,\mu^2\theta^2\bar{\theta}^2)\phi \tag{1.54}$$

where μ has the dimension of a mass (η must be changed accordingly). The substitution of (1.54) in $\bar{\Gamma}$ [according to (1.29)] has the effect of changing in particular the above IR-singular propagator into $1/(k^2-\mu^2)^2$. Thus μ^2 is an IR-cut-off.

Moreover, this IR-cut-off appearing as a parameter of the field redefinition, is a gauge parameter, in the same sense as the parameters a_k previously discussed. The proof is the same, too, although the presence of fields staying massless (but with non-singular propagators) complicates considerably the technical task of solving the BRS-cohomology.

Thus the physical quantities do not depend on the IR-cut-off. In other words, the IR-singularities cancel when computing these quantities.

Of course, the ansatz (1.54) breaks supersymmetry explicitly. This soft breaking is, however, controllable and can be shown not to affect the physical quantities.

II. CONFORMAL INVARIANCE[12),13)]

Let us consider the super-Yang-Mills model of Section I, with a simple gauge group G, and couple it with chiral matter fields A^R. Here R labels both the field and the irreducible representation of G where it lives. Its BRS transformation is

$$\begin{aligned} sA^R &= -c_{+i}\, T^i{}_R\, A^R \\ s\bar{A}_R &= \bar{A}_R\, T^i{}_R\, \bar{c}_+, \end{aligned} \tag{2.1}$$

the Hermitian matrices $T^i{}_R$ representing the generators of G in the representation R. The Slavnov identity (1.7) has a corresponding piece:

$$S(\Gamma\} = \ldots + \sum_{\ell} \int dS \frac{\delta\Gamma}{SY_R} \frac{\delta\Gamma}{\delta A^R} - c.c. \sim 0 \tag{2.2}$$

where Y_R is the external field coupled to the BRS transformation of A^R. The matter field contribution to the action (1.4) is

$$\Gamma_{matter} = 1/16 \int dV \sum_R \bar{A}_R \, e^{\phi_i T^i}{}^R A^R + \int dS \sum_I \lambda_I W^I(A) + Y_R sA^R] + c.c. \tag{2.3}$$

where $W^I(A)$ is a basis of invariant cubic polynomials of A, and the λ_I's are the "Yukawa" coupling constants. Mass terms, not necessarily gauge invariant but supersymmetric, are supposed to be present in order to avoid the off-shell infra-red problem discussed in Section I.3. The Slavnov identity is thus softly broken, as well as all other following equations [this is expressed by the sign $\sim$ in (2.2)].

Our strategy for studying the properties of the Callan-Symanzik β functions and for finding conditions under which they vanish is based on the existence of a BRS-invariant supercurrent[3)] V_μ obeying the Ward identity

$$\bar{D}^{\dot\alpha} V_{\alpha\dot\alpha} \sim -2w_\alpha \Gamma - 4/3 \, D_\alpha(S+S_0) \tag{2.4}$$

with

$$V_{\alpha\dot\alpha} = \sigma^\mu_{\alpha\dot\alpha} V_\mu, \quad \bar{D}_{\dot\alpha} S = \bar{D}_{\dot\alpha} S_0 = 0.$$

w_α is a functional differential operator expressing the different symmetries (superconformal group) involved. The letters V, S and S_0 stand for insertions[18)]. V and S are BRS-invariant, i.e.,

$$B_{\bar\Gamma} V_\mu \sim 0, \quad B_{\bar\Gamma} S \sim 0 \tag{2.5}$$

[see (1.16) and Ref. 18) for the definition of $B_{\bar\Gamma}$].

Let us forget S_0, an effect of the gauge fixing, irrelevant for the present discussion. The supercurrent V_μ contains[20)] among its components an axial current ($\theta = 0$ component) associated with R-invariance[21),3)] and the conserved symmetric energy-momentum tensor ($\theta\sigma\bar\theta$-component). The BRS-invariant chiral insertion S, of dimension 3, and of order $\hbar$, describes in particular the anomalies of the axial current and of the trace of the energy-momentum tensor, the latter being related to the dilatation anomaly, hence to the Callan-Symanzik equation. The precise relation is the following. We expand S in a basis of BRS-invariant insertions L_n of dimension

3 defined through the action principle[18)] by

$$\nabla_n \Gamma \sim \int dS L_n + \int d\bar{S} \bar{L}_n \tag{2.6}$$

where the ∇_n's are a basis of BRS-invariant differential operators:

$$\begin{aligned}
&\nabla_g = \partial_g, \ \nabla_{\lambda_I} = \partial_{\lambda_I}, \ \nabla_k = \partial_{a_k}, \\
&\nabla_\phi = \tilde{N}_\phi =: N_\phi - N_\rho - N_{c_-} - N_{\bar{c}_-} - N_B - N_{\bar{B}} + 2\alpha\partial_\alpha \\
&\nabla^R{}_S = \tilde{N}^R{}_S =: N_A{}^R{}_S - N_Y{}^R{}_S \\
&\nabla_+ = \tilde{N}_+ =: N_{c_+} - N_\sigma + \text{c.c.}
\end{aligned} \tag{2.7}$$

the N's being the counting operators

$$N_\phi = \int dV \, \phi \frac{\delta}{\delta\phi} , \quad N^R{}_S = \int dS A^R \frac{\delta}{\delta A^S} , \quad \text{etc.} \tag{2.8}$$

One sees in particular that

$$\begin{aligned}
L^R{}_S &= \left(A^R \frac{\delta}{\delta A^S} - Y_S \frac{\delta}{\delta Y_R}\right) \Gamma \\
L_+ &= \operatorname{Tr}\left(c_+ \frac{\delta}{\delta c_+} - \sigma \frac{\delta}{\delta\sigma}\right) \Gamma
\end{aligned} \tag{2.9}$$

and one can show that

$$L_\phi = \overline{DD} \, \ell_\phi, \quad L_k = \overline{DD} \, \ell_k \tag{2.10}$$

where ℓ_ϕ and ℓ_k are BRS-invariant.

We thus write

$$S = \beta_g L_g + \sum_I \beta_{\lambda_I} L_{\lambda_I} - \gamma_\phi L_\phi - \sum_{R,S} \gamma^S{}_R L^R{}_S - \gamma_+ L_+ - \sum_k \beta_k L_k . \tag{2.11}$$

The connection with the Callan-Symanzik equation

$$C\Gamma =: [\sum_a m_a \partial_{m_a} + \beta_g \partial_g + \beta_{\lambda_I} \partial_{\lambda_I} + \beta_{\bar{\lambda}_I} \partial_{\bar{\lambda}_I} - \gamma_\phi \tilde{N}_\phi - \gamma^S{}_R \tilde{N}^R{}_S - \gamma_+ \tilde{N}_+ - \qquad (2.12)$$
$$- \gamma_k \partial_{a_k}]\Gamma \sim 0$$

follows from the identity[3)]

$$\sum_a m_a \partial_{m_a} \Gamma + \int dSS + c.c. = 0 \qquad (2.13)$$

(summation is over all mass parameters of the theory).

Let us perform a change of basis for the insertions $L^R{}_S$, the new basis $\{L_{0a}, L_{1A}\}$ being defined according to (2.6) through the counting operators

$$\tilde{N}_{0a} = \sum_{R,S} e_a{}^S{}_R \tilde{N}^R{}_S$$
$$\tilde{N}_{1A} = \sum_{R,S} f_A{}^S{}_R \tilde{N}^R{}_S \qquad (2.14)$$

where the operators $\tilde{N}_{0a}$ form a basis of counting operators annihilating the superpotential terms W^I of the action (2.3):

$$\tilde{N}_{0a} W^I(A) = 0, \ \forall I \qquad (2.15)$$

The $\tilde{N}_{1A}$ complete the basis of matter field counting operators.

The expansion of S in this new basis reads [with (2.10) taken into account]

$$S = \beta_g L_g + \sum_I \beta_{\lambda_I} L_{\lambda_I} - \gamma_\phi \overline{DD}\, \ell_\phi - \sum_a \gamma_{0a} L_{0a} - \sum_A \gamma_{1A} L_{1A} - \gamma_+ L_+ - \sum_k \gamma_k \overline{DD}\, \ell_k . \qquad (2.16)$$

A key remark, now, is that the conditions (2.15) defining the insertion L_{0a} express the invariance of the theory (at the classical approximation) under the chiral transformations

$$\delta_a A^R = i\, e_a{}^R{}_S A^S, \quad \delta_a Y_R = -i\, e_a{}^S{}_R Y_S \qquad (2.17)$$

(which obviously leave invariant the rest of the action). This invariance can be extended to the quantum theory, but the associated axial currents become anomalous. A relation between the coefficients β, γ of (2.16) and the coefficients of these anomalies, as well as with the coefficient of the anomaly for the axial R-current will follow from the fact[12] that the BRS-invariant chiral insertion $T = S, L_g, \ldots,$ can be written in the form

$$T \sim \overline{DD}\,[rK^\circ + J^{inv}] + T^c \tag{2.18}$$

where T^c is genuinely chiral [i.e., it cannot locally be written as $\overline{DD}(\ldots)$] and J^{inv} is BRS-invariant. The insertion K° is not invariant, although $\overline{DD}K^\circ$ is, and is defined*) through the supersymmetric descent equations

$$\begin{aligned} &B_{\bar{\Gamma}}\, K^\circ \sim \bar{D}_{\dot\alpha}\, K^{1\dot\alpha} \qquad && B_{\bar{\Gamma}}\, K^{1\dot\alpha} \sim (\bar{D}D + 2\bar{D}D)^{\dot\alpha\alpha}\, K^2_\alpha \\ &B_{\bar{\Gamma}}\, K^2_\alpha \sim D_\alpha\, K^3 \qquad && B_{\bar{\Gamma}}\, K^3 \sim 0, \quad \bar{D}_{\dot\alpha}\, K^3 = 0 \end{aligned} \tag{2.19}$$

(the superscript denotes the ghost number).

The dimensionless insertion K^3 is proportional to $c^3{}_+$ and can be shown to be finite, hence uniquely defined up to a numerical factor, chosen to be 1/3 by convention. It turns out that K° is then uniquely defined up to an invariant. The coefficient r in (2.18) is thus defined unambiguously and is moreover gauge independent.

The genuinely chiral insertion T^c being expanded in terms of the chiral insertions L_{1A}, L_+, we can write (2.18), for $T = S, L_g, \ldots,$ as

$$\begin{aligned} &S \sim \overline{DD}[rK^\circ + J^{inv}] + S^c \qquad && L_g \sim \overline{DD}[(\frac{1}{128\, g^3} + r_g)K^\circ + J_g^{inv}] + L_g^c \\ &L_{\lambda_I} \sim \overline{DD}[r_{\lambda_I} K^\circ + J_I^{inv}] + L_I^c \qquad && L_{0a} \sim \overline{DD}[r_{0a}K^\circ + J_{0a}^{inv}] + L_{0a}^c . \end{aligned} \tag{2.20}$$

The corresponding expressions for L_ϕ, L_k, L_{1A}, L_+ have a coefficient $r = 0$ due to our choice of basis. The coefficients r, r_{λ_I}, r_g and r_{0a} are of order $\hbar$ at least. The zeroth order coefficient in L_g comes from the fact that

$$\overline{DD}K^\circ = \mathrm{Tr}\; F^\alpha F_\alpha + O(\hbar) \tag{2.21}$$

is the integrand of the Yang-Mills action (1.4).

---

*) Up to an invariant.

Moreover, the coefficients r and r_{0a} in (2.20), which can be interpreted as the anomalies of the axial currents associated with R-invariance and the chiral invariances (2.17), respectively, can be proved*) to be non-renormalized: they have only one-loop contributions, which may be computed, with the result

$$r = \frac{1}{512(4\pi)^2}(-3C_2(G) + \sum_R T(R))$$
$$r_{0a} = -\frac{1}{256(4\pi)^2}\sum_R e_a{}^R{}_R\, T(R). \tag{2.22}$$

T(R) is defined for the irreducible representation R by

$$T(R)\delta^{ij} = \mathrm{Tr}\; T^i{}_R\, T^j{}_R \tag{2.23}$$

and

$$C_2(G) = T(ad) \tag{2.24}$$

is the quadratic Casimir operator of the group. The numbers $e_a{}^R{}_R$ were defined by (2.14) and (2.15).

The substitution of the expressions (2.20) in (2.16) and the identification of the coefficients of K° yield the equation

$$r = \beta_g(\frac{1}{128g^3} + r_g) + \sum_I \beta_{\lambda_I} r_{\lambda_I} - \sum_a r_{0a}\gamma_{0a}. \tag{2.25}$$

This is the announced relation between the Callan-Symanzik functions and the anomaly coefficients r, r_{0a}. One sees in particular that r is proportional to the one-loop β_g function.

If the representations R of the matter fields are chosen such that the coefficients (2.22) vanish,

$$r = r_{0a} = 0 \tag{2.26}$$

then Eq. (2.25) becomes homogeneous in β_g, β_{λ_I}.

---

*) The proof[12] is based on: (i) the finiteness of $K^3 = 1/3\ \mathrm{Tr}\ c^3{}_+$; (ii) the Callan-Symanzik equation, obeyed by the superspace integrals of the insertion S and L_{0a} without any of their own anomalous dimensions.

If, moreover, the theory can be completely reduced[14], i.e., that all Yukawa coupling constants λ_I can be chosen as power series in the gauge coupling constant g, then these functions $\lambda_I(g)$ must be solutions of the reduction equations[14]

$$\beta_{\lambda_I} = \beta_g \partial_g \lambda_I \tag{2.27}$$

and Eq. (2.25) becomes

$$0 = \beta_g\left(\frac{1}{128g^3} + r_g + r_{\lambda_I} \partial_g \lambda_I\right) \tag{2.28}$$

whose solution is, in perturbation theory,

$$\beta_g = 0. \tag{2.29}$$

This also implies the vanishing of all β_{λ_I}, due to (2.27).

We have thus proved that the model is asymptotically scale invariant, as announced in the Introduction, if, first, the representations are such that the vanishing of the quantities (2.22) holds and, second, that the reduction equations (2.27) admit non-trivial solutions. The latter is, as a rule, true if it is verified in the one-loop approximation.

A closer look at the superconformal group reveals that the physical quantities are also superconformally covariant.

REFERENCES

1) P. Breitenlohner and D. Maison - These Proceedings.

2) S.J. Gates, M.T. Grisaru, M. Roček and W. Siegel - "Superspace", Benjamin/Cummings (1983).

3) O. Piguet and K. Sibold - "Renormalized Supersymmetry", Birkhaüser, Boston (1986).

4) O. Piguet and K. Sibold - Nucl.Phys. B197 (1982) 257; B248 (1984) 301.

5) G. Bonneau - These Proceedings.

6) K.S. Stelle - These Proceedings.

7) A. Blasi and R. Collina - These Proceedings.

8) J.W. Juer and D. Storey - Phys.Lett. B119 (1982) 125; Nucl.Phys. B216 (1983) 185.

9) O. Piguet and K. Sibold - Nucl.Phys. B253 (1985) 517.

10) H. Kluberg-Stern and J.B. Zuber - Phys.Rev. D12 (1975) 467, 482, 3159.

11) O. Piguet and K. Sibold - Nucl.Phys. B248 (1984) 336; B249 (1984) 396.

12) O. Piguet and K. Sibold - Phys.Lett. B177 (1986) 373; Int.J.Mod.Phys. A1 (1986) 913.

13) O. Piguet and K. Sibold - Conference "Renormalization Group-86", JINR, Dubna (1986).

14) W. Zimmermann - Commun.Math.Phys. 97 (1985) 211;
R. Oehme, K. Sibold and W. Zimmermann - Phys.Lett. 153B (1985) 142.

15) A.J. Parkes and P.C. West - Phys.Lett. B138 (1984) 99; Nucl.Phys. B256 (1985) 340.

16) S. Mandelstam - Nucl.Phys. B213 (1983) 149;
L. Brink, O. Lindgren and B. Nilsson - Nucl.Phys. B212 (1983) 401.

17) P.S. Howe, K.S. Stelle and P. West - Phys.Lett. 124B (1983) 55;
P.S. Howe, K.S. Stelle and P.K. Townsend - Nucl.Phys. B236 (1984) 125.

18) K. Sibold - These Proceedings.

19) O. Piguet and K. Sibold - Nucl.Phys. B247 (1984) 484.

20) S. Ferrara and B. Zumino - Nucl.Phys. B87 (1975) 207.

21) P. Fayet - Nucl.Phys. B90 (1975) 104.

$N = 2$ Supersymmetric Yang-Mills Theories in the Wess-Zumino Gauge

PETER BREITENLOHNER

Max-Planck-Institut für Physik und Astrophysik
– Werner-Heisenberg-Institut für Physik –
P.O.Box 40 12 12, Munich (Fed. Rep. Germany)

1. INTRODUCTION

We consider the renormalization of the $N = 2$ Yang-Mills multiplet coupled to matter and of the $N = 4$ Yang-Mills multiplet in the Wess-Zumino gauge. Since there is no formulation of the $N = 4$ theory with auxiliary fields we have to describe it as a $N = 2$ theory with one matter multiplet in the adjoint representation and have to impose additional constraints later on in order to guarantee the $N = 4$ supersymmetry. In view of the fact that all known consistent renormalization schemes violate either supersymmetry or gauge invariance, we study the possible anomalous radiative corrections to both the BRS and SUSY Ward identities.

The analogous program for the $N = 1$ theory has been performed in the very elegant superfield formulation, i.e. with unconstrained multiplets. This approach has the problem that there are massless scalar fields of canonical dimension zero. These not only require an IR-regulator destroying explicit BRS (Slavnov) or SUSY invariance, but also open the door for an infinite parameter group of field redefinitions [1]. Furthermore for extended supersymmetry there does not seem to exist an acceptable supermultiplet to put the Faddeev-Popov ghosts in. Last not least there is no superfield version of the $N = 4$ theory available. All these problems are avoided using the so-called Wess-Zumino gauge [2]. Yet, there is a price to pay:

i) the supersymmetry variations are non-linear;

ii) the commutator of two supersymmetry transformations contains a covariant translation instead of an ordinary, field independent one;

iii) the gauge fixing term violates supersymmetry explicitly.

We find, however, that this price is low compared to the trouble one avoids. Hence we shall use the Wess-Zumino gauge.

In this article we will concentrate on the non-linearity of the supersymmetry transformations in the Wess-Zumino gauge. This non-linearity originates from the non-linearity of the 'field dependent gauge transformations' and as a first step we will analyze the complications due to this non-linearity.

2. Invariant and Non-invariant Regularization Schemes

In this section we want to discuss in quite general terms the situation if some symmetry of the classical theory is or is not explicitly preserved by the regularization and renormalization procedure. In addition we want to recall that a renormalization procedure which deserves that name is more than a prescription to obtain finite results from divergent expressions. A renormalization procedure must in addition satisfy Hepp's axioms [3] which are equivalent to those locality and causality requirements which are the starting point to construct the perturbation expansion.

2.1. Invariant regularization schemes

If an invariance of the classical theory is explicitly preserved by the regularization and renormalization procedure the resulting renormalized theory will certainly be invariant. This very simple fact has motivated the invention of various regularization and renormalization procedures which preserve one or the other type of symmetry.

There are renormalization procedures which explicitly preserve supersymmetry, but none of them preserves gauge invariance. Conventional dimensional renormalization [4, 5] clearly violates supersymmetry because the structure of supersymmetry multiplets is different in different space-time dimensions. The so-called 'regularization by dimensional reduction' [6] method was soon found to be inherently inconsistent by its very inventor [7]. In spite of this it seems still to be quite popular [8]. The method of 'higher covariant derivatives' [9] either breaks gauge invariance [10] or does not regularize one-loop diagrams and has to be supplemented by another regularization breaking one of the desired invariances. A systematic study of such hybrid regularizations seems to be missing. Moreover it is not clear whether this method can be extended to $N = 2$ supersymmetry.

Given the fact that there is no acceptable renormalization scheme which explicitly preserves gauge (or rather BRS) invariance and supersymmetry we have to study the consequences of violations of supersymmetry (and possibly other symmetries) by the process of renormalization.

2.2. Non-invariant regularization schemes

It is a fact, although not a widely recognized one, that the existence of an invariant regularization with respect to some desired symmetry is only of marginal interest from a more general standpoint. In fact, since the pioneering work of Becchi, Rouet and Stora [11] on the renormalization of gauge theories it is understood that the problem of genuine anomalies in Ward-Takahashi identities is a purely algebraic one. It can be reduced to the cohomology theory of Lie algebras. Looking at the problem from this more general standpoint one has freed oneself from the necessity to refer to any particular renormalization scheme. We will only assume that Lorentz invariance and invariance under global compact groups (in our case $SU(2) \times SU(2)$) are preserved. This is no loss of generality because it is known that these symmetries can always be restored (absence of anomalies for these symmetries).

The classical Lagrangean is highly restricted (relations between coefficients and absence of certain terms) by symmetry requirements. If a symmetry is destroyed by the regularization procedure, there is no more reason for these restrictions in the regularized Lagrangean. Such a restriction has in fact no meaning independent of a particular renormalization scheme. We must, therefore, start from a more general 'effective' Lagrangean containing all possible terms with arbitrary coefficients with no other restrictions than those imposed by power counting and by symmetries which are respected by the renormalization procedure.

The invariance or non-invariance of the renormalized theory under the desired symmetries is most easily expressed by the absence or presence of anomalous terms in the corresponding Ward-Takahashi identities. One must try to adjust the many additional parameters in the effective action in such a way that all anomalies are removed. The resulting symmetric renormalized theory should then have as many free parameters as the original classical theory. They can be fixed by suitable symmetric normalization conditions, such that the resulting theory is completely determined by symmetry requirements and normalizations conditions independent of any particular scheme.

3. Ward Identities and Wess-Zumino Consistency Conditions

Invariances of the Lagrangean are reflected at the level of generating functionals by Ward identities. In this section we want to derive these Ward identities and in particular point out the difference between linear and non-linear transformations of the elementary fields $\phi = \begin{pmatrix} \phi_1 \\ \vdots \end{pmatrix}$.

3.1. Linear transformations of the fields

Let us consider a set of (infinitesimal) transformations δ_i of the fields ϕ which leave the Lagrangean invariant

$$\delta_i \mathcal{L}_{inv} = 0 \tag{3.1}$$

and satisfy the commutation relations of some Lie algebra

$$[\delta_i, \delta_j] = f_{ij}{}^k \delta_k \,. \tag{3.2}$$

In order to simplify the following discussion we ignore (for the moment) all sign factors due to Fermi fields or supersymmetry transformations.

If all the fields transform linearly we have

$$\delta_i \phi = t_i \phi \tag{3.3}$$

with some matrix representation t_i of the Lie algebra. In order to generate Green's functions we have to use the 'classical' Lagrangean obtained by adding source terms

$$\mathcal{L}_{cl} = \mathcal{L}_{inv} + j^T \phi \,. \tag{3.4}$$

This classical Lagrangean satisfies the identities

$$(\delta_i + W_i)\mathcal{L}_{cl} \equiv 0 \tag{3.5}$$

where the W_i's are differential operators in the sources

$$W_i = -j^T t_i \frac{\delta}{\delta j^T} \tag{3.6}$$

which satisfy by construction the commutation relations

$$[W_i, W_j] = f_{ij}{}^k W_k \,. \tag{3.7}$$

The naive action principle (valid for the tree approximation) implies that the generating functional $Z(j)$ of the connected Green's functions satisfies the naive Ward identities

$$W_i Z(j) = 0 \,. \tag{3.8}$$

These relations will, in general, not be true for the renormalized theory. The renormalized action principle [12, 13, 14, 5], which is a consequence of general results from renormalization theory, yields

$$W_i Z = \Delta_i Z \tag{3.9}$$

where the anomalies Δ_i are integrated local operator insertions of appropriate dimension and covariance which are at least of order $O(\hbar)$, i.e. vanish in the tree approximation. In order to study how the Δ_i's change if we perform finite renormalizations (i.e. exploit the freedom inherent to any renormalization scheme) we have to introduce the generating functional $\Gamma(\phi)$ of 1PI Green's functions (vertex functions) obtained from Z by a Legendre transformation with respect to the sources j. We first define

$$\phi^T(j) = \frac{\delta Z}{\delta j} \tag{3.10}$$

solve for $j(\phi)$ and define

$$\Gamma(\phi) = Z - j^T \phi \tag{3.11}$$

with the consequence

$$j^T(\phi) = -\frac{\delta \Gamma}{\delta \phi} \,. \tag{3.12}$$

Note that the tree approximation Γ_{cl} of Γ coincides with $\mathcal{L}_{inv}$

$$\Gamma = \Gamma_{cl} + O(\hbar)\,, \qquad \Gamma_{cl} = \mathcal{L}_{inv}\,. \tag{3.13}$$

Upon Legendre transformation the Ward identities take the form

$$W_i(\Gamma) = \Delta_i \Gamma \tag{3.14}$$

where the differential operator

$$W_i(\Gamma) \equiv \phi^T t_i^T \frac{\delta \Gamma}{\delta \phi^T} \tag{3.15}$$

is *linear* in Γ. As a consequence of the commutation relations (7) the anomalies Δ_i satisfy the Wess-Zumino consistency conditions [15]

$$W_i^\Gamma(\Delta_j \Gamma) - W_j^\Gamma(\Delta_i \Gamma) = f_{ij}{}^k \Delta_k \Gamma \tag{3.16}$$

where

$$W_i^\Gamma X \equiv \phi^T t_i^T \frac{\delta X}{\delta \phi^T} \equiv W_i(X) \tag{3.17}$$

does not depend on Γ.

3.2. Non-linear transformations of the fields

Let us now consider the case where the variations of the elementary fields are some non-linear expressions P_i in the fields

$$\delta_i \phi = P_i(\phi) \tag{3.18}$$

and consequently

$$\delta_i(\mathcal{L}_{inv} + j^T \phi) = j^T P_i(\phi) \, . \tag{3.19}$$

In order to express these changes through differential operators W_i acting on sources we have to introduce additional sources for all non-linear expressions (composite fields) appearing in the P_i's as well as for all their iterated variations. Let $\Phi = \begin{pmatrix} \phi \\ \phi_c \end{pmatrix}$ be all these fields, ϕ are the elementary ones as before and ϕ_c are all the composite fields (possibly infinitely many), such that the infinitesimal variations are again linear (in Φ)

$$\delta_i \Phi = t_i \Phi \, . \tag{3.3'}$$

Similarly we introduce sources $J = \begin{pmatrix} j \\ j_c \end{pmatrix}$ and proceed as before: the differential operators

$$W_i = -J^T t_i \frac{\delta}{\delta J^T} \equiv -(\, j^T \quad j_c^t \,)\, t_i \begin{pmatrix} \frac{\delta}{\delta j^T} \\ \frac{\delta}{\delta j_c^T} \end{pmatrix} \tag{3.6'}$$

act on the generating functional $Z(J) \equiv Z(j, j_c)$.

The vertex functional Γ is obtained from Z by a Legendre transformation with respect to the sources j for elementary fields but *not* the sources j_c for the composite fields. We define

$$\phi^T(j, j_c) = \frac{\delta Z}{\delta j} \tag{3.10'}$$

solve for $j(\phi, j_c)$ and define

$$\Gamma(\phi, j_c) = Z - j^T \phi \tag{3.11'}$$

with the consequence

$$j^T(\phi, j_c) = -\frac{\delta \Gamma}{\delta \phi} \, . \tag{3.12'}$$

In the tree approximation we find

$$\Gamma_{cl} = \mathcal{L}_{inv} + j_c^T \phi_c \, . \tag{3.13'}$$

The differential operator

$$W_i(\Gamma) \equiv (\, \phi^T \quad \tfrac{\delta \Gamma}{\delta j_c} \,)\, t_i^T \begin{pmatrix} \frac{\delta \Gamma}{\delta \phi^T} \\ -j_c \end{pmatrix} \tag{3.15'}$$

is now *non-linear* in Γ and

$$W_i^\Gamma X \equiv (\,\phi^T \quad \tfrac{\delta\Gamma}{\delta j_c}\,)\, t_i^T \begin{pmatrix} \frac{\delta X}{\delta\phi^T} \\ 0 \end{pmatrix} + (\,0 \quad \tfrac{\delta X}{\delta j_c}\,)\, t_i^T \begin{pmatrix} \frac{\delta\Gamma}{\delta\phi^T} \\ -j_c \end{pmatrix} \not\equiv W_i(X) \qquad (3.17')$$

does depend on Γ.

3.3. The cohomology problem

We can now use the fact

$$\Delta_i\Gamma = \Delta_i + O(\hbar\Delta_i)\,, \qquad (3.20)$$

assume

$$\Delta_i = \hbar^m \Delta_i^{(m)} + o(\hbar^m)\,, \qquad (3.21)$$

and use the consistency conditions (16) to the order $\hbar^m$ to obtain the cohomology equations

$$W_i^{cl}\Delta_j^{(m)} - W_j^{cl}\Delta_i^{(m)} = f_{ij}{}^k\Delta_k^{(m)} \qquad (3.22)$$

where W_i^{cl} is given by eq. (17) resp. (17′) with the replacement $\Gamma \to \Gamma_{cl}$. The difference between the two cases of linear and non-linear transformations is that the structure of W_i^{cl} is much simpler in the former case.

If we change the effective action (which is in a certain sense the renormalized version of $\mathcal{L}_{inv}$ or rather Γ_{cl})

$$\Gamma_{eff} \to \Gamma_{eff} + \hbar^m\Delta_{\mathcal{L}}^{(m)} \qquad (3.23)$$

the corresponding change in the anomalies is

$$\Delta_i^{(m)} \to \Delta_i^{(m)} + W_i^{cl}\Delta_{\mathcal{L}}^{(m)}\,. \qquad (3.24)$$

This poses the following *cohomology problem*: Given some anomalies $\Delta_i^{(m)}$ which necessarily satisfy the consistency conditions (22). If there exists some integrated local expression X such that

$$\Delta_i^{(m)} = W_i^{cl}X \qquad (3.25)$$

then the change $\Gamma_{eff} \to \Gamma_{eff} - \hbar^m X$ will remove all anomalies in this order in $\hbar$, otherwise there is a genuine anomaly.

The procedure described above is the standard procedure to establish BRS invariance. The main difficulty arises from the fact that the differential operators W_i^{cl} mix terms with a different number of sources. The situation is, however, not too bad for BRS transformations because they are nilpotent and therefore they require only a finite number of j_c's. For SUSY in the Wess-Zumino gauge the situation is more complicated because there is an infinite number of j_c's and moreover almost all of them have negative (power counting) dimension.

4. $N = 2$ Super Yang-Mills with Matter

We use a notation with $SU(2)$ covariant $N = 2$ Majorana spinors and assume that the gauge group is simple and non-chiral and that all matter fields are massless. These theories have a global $SU(2) \times SU(2)$ invariance but only the diagonal $SU(2)$ subgroup is made explicit by our notation. The fields of these theories are contained in one $N = 2$ Yang-Mills multiplet [16]

$$(A_a,\ S,\ P;\ \lambda;\ \vec{D}) \tag{4.1}$$

in the adjoint representation of the Lie algebra (or equivalently with values in the Lie algebra) as well as one $N = 2$ matter multiplet (hypermultiplet) [16]

$$(A,\ \vec{A};\ \alpha;\ F,\ \vec{F}) \tag{4.2}$$

transforming under an arbitrary representation ρ of the Lie algebra. This representation need not be irreducible but we assume it to be real.

The Lagrangean has the form $\mathcal{L}_{inv} = \mathcal{L}_{YM} + \mathcal{L}_M$ where

$$\begin{aligned}\mathcal{L}_{YM} = \frac{1}{g^2}\Big(&-\frac{1}{4}F^{ab}\cdot F_{ab} + \frac{1}{2}D^aS\cdot D_aS + \frac{1}{2}D^aP\cdot D_aP - \frac{i}{2}\bar{\lambda}\cdot\gamma^a D_a\lambda \\ &+\frac{1}{2}\vec{D}\cdot\vec{D} - \frac{i}{2}\bar{\lambda}\cdot(S - i\gamma_5 P)\times\lambda - \frac{1}{2}(S\times P)\cdot(S\times P)\Big)\end{aligned} \tag{4.3}$$

and

$$\begin{aligned}\mathcal{L}_M = &\frac{1}{2}D^aA\cdot D_aA + \frac{1}{2}D^a\vec{A}\cdot D_a\vec{A} - \frac{i}{2}\bar{\alpha}\cdot\gamma^a D_a\alpha + \frac{1}{2}F\cdot F \\ &+\frac{1}{2}\vec{F}\cdot\vec{F} + A\cdot\rho(\vec{D})\vec{A} + \frac{1}{2}\vec{A}\cdot\rho(\vec{D})\times\vec{A} - i\rho(\bar{\lambda})(A + i\vec{\tau}\vec{A})\cdot\alpha \\ &+\frac{i}{2}\bar{\alpha}\cdot\rho(S + i\gamma_5 P)\alpha + \frac{1}{2}A\cdot\big(\rho(S)\rho(S) + \rho(P)\rho(P)\big)A \\ &+\frac{1}{2}\vec{A}\cdot\big(\rho(S)\rho(S) + \rho(P)\rho(P)\big)\vec{A}\,.\end{aligned} \tag{4.4}$$

The supersymmetry variations $\delta(i\bar{\epsilon}Q)\phi$ generate the $N = 2$ SUSY algebra [17]

$$\frac{1}{2}[\delta(i\bar{\epsilon}_1 Q), \delta(i\bar{\epsilon}_2 Q)] = -i\bar{\epsilon}_1\epsilon_2\big(\delta(S) + \delta(Z)\big) - \bar{\epsilon}_1\gamma_5\epsilon_2\delta(P) - i\bar{\epsilon}_1\gamma^a\epsilon_2\delta(T_a) \tag{4.5}$$

where $\delta(Z)$ is the central charge transformation (acting on the matter fields only), $\delta(T_a) \equiv \partial_a - \delta(A_a)$ is the gauge covariant translation, and $\delta(A_a)$, $\delta(S)$, ... are field dependent gauge transformations. These field dependent gauge transformations are the source of all non-linearities in the transformation laws and force us to introduce an infinite number of composite fields with increasing dimensions.

If we define a spinor derivative $\mathcal{D}$ by

$$\delta(i\bar{\epsilon}Q)\phi = i\bar{\epsilon}\mathcal{D}\phi \tag{4.6}$$

we can rewrite the SUSY algebra in the form

$$\begin{aligned}
\bar{\mathcal{D}}\mathcal{D} &= -8i\big(\delta(S) + \delta(Z)\big) \\
\bar{\mathcal{D}}\gamma_5\mathcal{D} &= -8\delta(P) \\
\bar{\mathcal{D}}\gamma_a\mathcal{D} &= -8i\delta(T_a) \\
\bar{\mathcal{D}}\gamma_a\gamma_5\vec{\tau}\mathcal{D} &= 0 \\
\bar{\mathcal{D}}\gamma_{ab}\vec{\tau}\mathcal{D} &= 0
\end{aligned} \tag{4.7}$$

and find in addition

$$[\mathcal{D}, \delta(Z)] = 0\,. \tag{4.8}$$

These commutation relations (7) generate an infinite dimensional algebra through the identities

$$\begin{aligned}
[\mathcal{D}, \delta(T_a)] &= -[\mathcal{D}, \delta(A_a)] = -\delta(\mathcal{D}A_a) \\
[\mathcal{D}, \delta(X)] &= \delta(\mathcal{D}X) \\
[\delta(T_a), \delta(T_b)] &= -\delta(F_{ab}) \\
[\delta(T_a), \delta(X)] &= \delta(D_a X)
\end{aligned} \tag{4.9}$$

where X can be any of the covariant fields S, P, … .

The BRS transformation acts in the usual way on the physical fields

$$sA_a = D_a\bar{c}, \quad sS = \bar{c} \times S, \quad \ldots \quad s\vec{F} = \rho(\bar{c})\vec{F} \tag{4.10}$$

and on the ghost fields c, $\bar{c}$ and B

$$sc = B, \quad sB = 0, \quad s\bar{c} = \frac{1}{2}\bar{c} \times \bar{c} \tag{4.11}$$

such that

$$\{s\,,\,s\} = 0\,. \tag{4.12}$$

The gauge fixing term

$$\begin{aligned}
\mathcal{L}_{\Phi\Pi} &= \frac{1}{\alpha g^2} s(c \cdot \partial_a A^a - \frac{1}{2}c \cdot B) \\
&= \frac{1}{\alpha g^2}(B \cdot \partial_a A^a - \frac{1}{2}B \cdot B - c \cdot \partial_a D^a \bar{c})
\end{aligned} \tag{4.13}$$

added to the Lagrangean is BRS invariant. Finally the ghost fields are invariant under SUSY transformations

$$\mathcal{D}c = \mathcal{D}\bar{c} = \mathcal{D}b = 0 \tag{4.14}$$

with the consequence

$$\{\mathcal{D}\,,\, s\} = 0\,. \tag{4.15}$$

With our definition (14) the gauge fixing term (13) is not SUSY invariant

$$\mathcal{D}\mathcal{L}_{\Phi\Pi} = \frac{1}{\alpha g^2} s(c\cdot\gamma^a\partial_a\lambda) \neq 0\,. \tag{4.16}$$

This leads to a slight complication because supersymmetry is already broken at the tree level. Due to the form $s(\ldots)$ the breaking term (16) does, however, not affect the physical sector (gauge invariant amplitudes). The following table collects all the relevant transformations together with some notation

δ_i	$\Delta_i^{(m)}$	W_i^{cl}
s	$\Delta_{\mathcal{S}}$	$\mathcal{S}$
$\mathcal{D}$	$\Delta_{\mathcal{W}}$	$\mathcal{W}$
$\delta(Z)$	Δ_Z	W_Z
$\delta(T_a)$	Δ_a	W_a
$\delta(X)$	Δ_X	W_X
$\tilde{\delta}(T_a)$	$\tilde{\Delta}_a$	$\tilde{W}_a$
$\tilde{\delta}(X)$	$\tilde{\Delta}_X$	$\tilde{W}_X$

(4.17)

where again X can be any of the covariant fields S, P, ... Since all insertions are integrated ones we have actually $\delta(T_a) = -\delta(A_a)$. The transformations $\tilde{\delta}(T_a)$ and $\tilde{\delta}(X)$ are not yet defined but will turn out to be useful later on.

Assume that all anomalies have been removed up to the order $\hbar^{m-1}$ for some $m \geq 1$, i.e. they all start at the order $\hbar^m$. Our aim is to show that the consistency conditions (3.22) imply that they can all be removed by a suitable change in Γ_{eff} (in this order $\hbar^m$). This will then define appropriately renormalized composite operators such that BRS and SUSY invariance are restored.

In the following sections this program will be performed for some of the symmetries under consideration.

5. BRS INVARIANCE

As a first step we want to study the BRS anomaly $\Delta_{\mathcal{S}}$ which must satisfy, due to the commutation relation (4.12), the usual consistency condition

$$\mathcal{S}\Delta_{\mathcal{S}} = 0\,. \tag{5.1}$$

In spite of the infinitely many sources and the complicated structure of the differential operator $\mathcal{S}$ it is relatively easy to show that any such anomaly has the structure

$$\Delta_{\mathcal{S}} = \mathcal{S}X_{\mathcal{S}} \tag{5.2}$$

with some $X_{\mathcal{S}}$ and can therefore be removed by a suitable finite renormalization.

Since there are infinitely many composite fields with increasing power counting dimension the dimension of their sources will soon get negative. Consequently the usual power counting arguments cannot be used to control the occurence of such sources, e.g., in $\Delta_{\mathcal{S}}$. We can, however, rearrange the sources $J = (j_e,\, J_c)$ as $J = (j^\alpha,\, k^\alpha,\, J^\alpha,\, K^\alpha)$ in the following way: The elementary fields $\varphi_\alpha = (A_a,\, S,\, P,\, \lambda,\, \vec{D},\, A,\, \vec{A},\, \alpha,\, F,\, \vec{F},\, c,\, \bar{c})$ are coupled to j^α and their BRS variations $s\varphi_\alpha$ to k^α. Note that the elementary field $B = sc$ is contained among $s\varphi_\alpha$. The 'physical' composite fields $\Phi_\alpha = (F_{ab}, \ldots)$, all with vanishing BRS charge, are coupled to J^α and their BRS variations to $s\Phi_\alpha$ to K^α.

We can decompose Γ_{cl} and $\mathcal{S}$ according to their degree in the sources J^α and K^α and find $\Gamma_{cl} = \Gamma_0 + \Gamma_1$, $\mathcal{S} = \mathcal{S}_0 + \mathcal{S}_1$ where (up to signs due to anti-commuting fields)

$$\Gamma_0 = \Gamma_{cl}|_{J=K=0} \tag{5.3a}$$

$$\Gamma_1 = J^\alpha \Phi_\alpha + K^\alpha s\Phi_\alpha \tag{5.3b}$$

$$\mathcal{S}_0 = s\varphi_\alpha \frac{\delta}{\delta\varphi_\alpha} + J^\alpha \frac{\delta}{\delta K^\alpha} + \frac{\delta\Gamma_0}{\delta\varphi_\alpha}\frac{\delta}{\delta k^\alpha} \tag{5.4a}$$

$$\mathcal{S}_1 = \Big(J^\beta \frac{\delta\Phi_\beta}{\delta\varphi_\alpha} + K^\beta \frac{\delta s\Phi_\beta}{\delta\varphi_\alpha}\Big)\frac{\delta}{\delta k^\alpha}\,. \tag{5.4b}$$

The first two terms in $\mathcal{S}_0$ are nice and simple, the third term involving field equations in the presence of sources k^α is somewhat messy and $\mathcal{S}_1$ contains the really unpleasant contribution of the infinite sequence of sources J^α and K^α to the field equations. The nilpotency of $\mathcal{S}$ implies

$$\mathcal{S}_0{}^2 = \big\{\mathcal{S}_0\,,\, \mathcal{S}_1\big\} = \mathcal{S}_1^2 = 0\,. \tag{5.5}$$

Similarly we express $\Delta_{\mathcal{S}}$ in the form

$$\Delta_{\mathcal{S}} = \sum_{n=0}^{\infty} \Delta_n \tag{5.6}$$

where each term Δ_n is homogeneous of degree n in the J'a and K's and obtain the consistency conditions

$$\begin{aligned} \mathcal{S}_0\Delta_0 &= 0 \\ \mathcal{S}_0\Delta_n + \mathcal{S}_1\Delta_{n-1} &= 0 \qquad \text{for} \quad n \geq 1\,. \end{aligned} \tag{5.1'}$$

It can be shown that the first of these equations implies that there exists an X_0 such that $\Delta_0 = \mathcal{S}_0 X_0$. Here we can use the original argument of [11] with minor modifications due to the presence of the elementary fields $\vec{D}$, F and $\vec{F}$ of dimension two. and we have to use the fact that the gauge group is non-chiral and therefore there can be no Adler-Bardeen anomaly.

We can now use the following recursive argument to show that the anomaly has indeed the form (2): Assume the first non-vanishing term in the sum (1′) has the form $\Delta_{n-1} = \mathcal{S}_0 X_{n-1}$ (true for $n = 1$). Subtracting this X_{n-1} from Γ_{eff} removes this term and modifies the next one Δ_n such that the consistency condition (1′) implies

$$\mathcal{S}_0\Delta_n = 0\,. \tag{5.7}$$

Using the fact that the anti-commutator

$$\left\{K^\alpha \frac{\delta}{\delta J^\alpha}, \mathcal{S}_0\right\} = J^\alpha \frac{\delta}{\delta J^\alpha} + K^\alpha \frac{\delta}{\delta K^\alpha} \tag{5.8}$$

yields the counting operator for the degree of homogeneity in the J's and K's, we find from eq. (7)

$$\Delta_n = \mathcal{S}_0 X_n \qquad \text{with} \qquad X_n = \frac{1}{n} K^\alpha \frac{\delta}{\delta J^\alpha}\Delta_n\,. \tag{5.9}$$

Thus we have shown that the anomaly $\Delta_{\mathcal{S}}$ can indeed be written in the form (2).

Next we must study the remaining finite renormalizations compatible with BRS invariance, i.e. the most general solution of the equation

$$\mathcal{S}\Delta_{\mathcal{L}} = 0 \tag{5.10}$$

where $\Delta_{\mathcal{L}}$ is an operator insertion of dimension four with vanishing ghost charge whereas $\Delta_{\mathcal{S}}$ had dimension five and ghost charge one. Repeating the arguments used to analyze $\Delta_{\mathcal{S}}$ we find

$$\Delta_{\mathcal{L}} = \mathcal{S}X_{\mathcal{L}} + \Delta_{\mathcal{L}}^{g.i.} \tag{5.11}$$

where $\Delta_{\mathcal{L}}^{g.i.}$ is a gauge invariant expression (without sources and ghost fields). Unfortunately the decomposition (11) is not unique, i.e. there exist some $X_{\mathcal{L}}$ such that $\mathcal{S}X_{\mathcal{L}}$ is a (non-vanishing) gauge invariant expression.

As a next step $X_{\mathcal{L}}$ and $\Delta_{\mathcal{L}}^{g.i.}$ could be chosen in such a way that (if possible) $\Delta_{\mathcal{W}}$ is removed. All higher anomalies Δ_Z, Δ_S, ... would then automatically vanish. This should then fix all the coefficients in Γ_{eff} except a few, i.e. the requirement of BRS and SUSY invariance should uniquely determine the theory up to a redefinition of the gauge coupling constant g and three (physically irrelevant) wave function renormalizations.

6. Field Dependent Gauge Transformations

6.1. SUSY Ward identities

Due to the commutation relations (4.7) we have the consistency conditions

$$\bar{\mathcal{W}}\Delta_{\mathcal{W}} = -8i\big(\Delta_S + \Delta_Z\big) \tag{6.1a}$$

$$\bar{\mathcal{W}}\gamma_5\Delta_{\mathcal{W}} = -8\Delta_P \tag{6.1b}$$

$$\bar{\mathcal{W}}\gamma_a\Delta_{\mathcal{W}} = -8i\Delta_a \tag{6.1c}$$

$$\bar{\mathcal{W}}\gamma_a\gamma_5\vec{\tau}\Delta_{\mathcal{W}} = 0 \tag{6.1d}$$

$$\bar{\mathcal{W}}\gamma_{ab}\vec{\tau}\Delta_{\mathcal{W}} = 0\,. \tag{6.1e}$$

If there were only the field dependent gauge transformations $\delta(S)$, $\delta(P)$, ..., they would generate an infinite dimensional and essentially free Lie algebra which would yield no useful consistency conditions. In the present case, where all field dependent gauge transformations are generated from supersymmetry transformations, we have additional consistency conditions. The supersymmetry transformations laws for the Yang-Mills multiplet

$$\mathcal{D}A_a = \gamma_a\lambda \tag{6.2a}$$

$$\mathcal{D}S = \lambda \tag{6.2b}$$

$$\mathcal{D}P = -i\gamma_5\lambda \tag{6.2c}$$

$$\mathcal{D}\bar{\lambda} = -i\gamma^a D_a(S + i\gamma_5 P) - \frac{i}{2}\gamma^{ab}F_{ab} + \gamma_5 S\times P - \vec{\tau}\vec{D} \tag{6.2d}$$

$$\mathcal{D}\vec{D} = i\vec{\tau}\big(-\gamma^a D_a\lambda - (S - i\gamma_5 P)\times\lambda\big) \tag{6.2e}$$

yield, together with the identities (4.8, 4.9),

$$\mathcal{W}(\Delta_Z + \Delta_S) - (W_Z + W_S)\Delta_{\mathcal{W}} = \Delta_\lambda \tag{6.3a}$$

$$\mathcal{W}\Delta_P - W_P\Delta_{\mathcal{W}} = -i\gamma_5\Delta_\lambda \tag{6.3b}$$

$$\mathcal{W}\Delta_a - W_a\Delta_{\mathcal{W}} = -\gamma_a\Delta_\lambda \tag{6.3c}$$

$$\mathcal{W}\bar{\Delta}_\lambda + W_\lambda\bar{\Delta}_{\mathcal{W}} = -i\gamma^a(W_a\Delta_S - W_S\Delta_a) + \ldots - \vec{\tau}\Delta_{\vec{D}} \tag{6.3d}$$

$$\mathcal{W}\Delta_{\vec{D}} - W_{\vec{D}}\Delta_{\mathcal{W}} = -i\gamma^a\vec{\tau}(W_a\Delta_\lambda - W_\lambda\Delta_a + \ldots)\,. \tag{6.3e}$$

Each of the anomalies $\Delta_{\mathcal{W}}$, Δ_S, ... is BRS invariant and has a decomposition into a piece (depending on sources and/or ghost fields) which is itself a BRS variation and a gauge invariant piece (compare eq. (5.11)) but again these decompositions are not unique.

6.2. A New Set of Transformations

The form of the consistency conditions (1, 3) would simplify considerably if we could first remove the anomalies Δ_a, Δ_S, ... for the field dependent gauge transformations. In spite of the fact that all these transformations essentially form a free Lie-algebra this can indeed be done. Note that all these transformations commute with s (again we ignore a sign change due to fermion fields e.g. in $\delta(\lambda)$). It should, therefore, not be too much of a surprise that each of them can be expressed as the anti-commutator of a new transformation with s. We thus introduce the new transformations $\tilde{\delta}(T_a)$ and $\tilde{\delta}(X)$ which act on the fields as follows (all physical fields are annihilated)

φ	$\tilde{\delta}(T_a)\varphi$	$\tilde{\delta}(X)\varphi$	$\partial_a\varphi - \{\tilde{\delta}(T_a), s\}\varphi$	$\{\tilde{\delta}(X), s\}\varphi$
c	0	0	0	0
B	$\partial_a c$	0	0	0
$\bar{c}$	A_a	X	0	0
A_b	0	0	F_{ab}	$D_b X$
Y	0	0	$D_a Y$	$\delta(X) Y$

(6.4)

an X and Y can be any of the covariant fields S, P, ...

These new transformations are linear and the corresponding Ward identity operators are rather simple. Moreover they are all nilpotent and mutually anti-commute. As a consequence the corresponding consistency conditions are extremely powerful. Once we have removed their anomalies $\tilde{\Delta}_a$ and $\tilde{\Delta}_X$ the anomalies Δ_a and Δ_X for the field dependent gauge transformations vanish automatically.

The Lagrangean $\mathcal{L}_{inv} = \mathcal{L}_{YM} + \mathcal{L}_M$ is obviously invariant under these new transformations but the gauge fixing term $\mathcal{L}_{\Phi\Pi}$ is not (compare eq. (4.17)). This explicit breaking which is already present at the tree level can be taken into account by additional terms in the Ward identity operators. In order to do so we have to add some more sources. The constant (space-time independent) sources $\tilde{J}^\alpha$ and $\tilde{K}^\alpha$ couple to integrated composite fields $\int \tilde{\Phi}_\alpha$ and $\int s\tilde{\Phi}_\alpha$ which contain the variations of $\mathcal{L}_{\Phi\Pi}$ under iterated application of all the transformations under consideration. Since we can treat these new sources $\tilde{J}$ and $\tilde{K}$ in almost the same way as the J's and K's all results about BRS invariance remain valid. Among the new (integrated) composite fields there are in particular $\int \tilde{\Phi}_a = \tilde{\delta}(T_a)\mathcal{L}_{\Phi\Pi}$ and $\int \tilde{\Phi}_X = \tilde{\delta}(X)\mathcal{L}_{\Phi\Pi}$ (for $X = S,\, P,\, \ldots$). The non-invariance of $\mathcal{L}_{\Phi\Pi}$ yields the inhomogeneous terms $-\frac{\partial}{\partial \tilde{J}_a}$ and $-\frac{\partial}{\partial \tilde{J}_X}$ in $\tilde{W}_a$ and $\tilde{W}_X$ respectively. For each such transformation, e.g. for $\tilde{\delta}(S)$, we can use the identity

$$\left\{\tilde{W}_S\,,\, -\tilde{J}_S\right\} = 1 \tag{6.5}$$

and find indeed that

$$\tilde{\Delta}_S = \tilde{W}_S \Delta_{\mathcal{L}} \qquad \text{with} \qquad \Delta_{\mathcal{L}} = -\tilde{J}^S \tilde{\Delta}_S\,. \tag{6.6}$$

Subtracting this term from Γ_{eff} removes this anomaly (to the order in $\hbar$ under consideration). The fact that this can be done should not be very surprising, in a sense we are just redefining the composite field $\tilde{\delta}(S)\mathcal{L}_{\Phi\Pi}$ which is the tree level value of this anomaly. Having done this we are left with modified values for all the other anomalies. In the next step we can remove one of the remaining anomalies, say $\tilde{\Delta}_P$ in exactly the same way. Moreover since $\tilde{J}^S$ anti-commutes with $\tilde{W}_P$ the $\tilde{\Delta}_S$ removed in the previous step stays absent. Repeating this process we can construct a $\Delta_{\mathcal{L}}$ such that

$$\begin{aligned} \tilde{\Delta}_a &= \tilde{W}_a \Delta_{\mathcal{L}}\,, \\ \tilde{\Delta}_X &= \tilde{W}_X \Delta_{\mathcal{L}} \qquad \text{for all} \quad X\,. \end{aligned} \tag{6.7}$$

This $\Delta_{\mathcal{L}}$ is not yet BRS-invariant and will, therefore, introduce a new BRS-anomaly $\Delta_{\mathcal{S}}$. In the process described above we have, however, introduced at least one explicit power of $\tilde{J}$, i.e., increased the number of J's and K's which played a crucial rôle in the removal of the BRS-anomaly $\Delta_{\mathcal{S}}$. In order to remove all the anomalies $\Delta_{\mathcal{S}}$, $\tilde{\Delta}_a$ and $\tilde{\Delta}_X$ simultaneously we have to use the process described above after each step of the inductive procedure used to remove $\Delta_{\mathcal{S}}$ (compare eq. (5.9)).

The anomalies $\Delta(T_a)$ and $\Delta(X)$ for the field dependent gauge transformations will now all vanish automatically. At the same time the vast majority of the parameters in the effective action has been determined. All the remaining freedom should now suffice to remove the SUSY-anomalies.

7. Conclusions

We have shown that one can study the renormalization of supersymmetric Yang-Mills theories in a way which does not depend on any particular renormalization scheme. Since there is no known reliable renormalization procedure which respects both BRS invariance and supersymmetry we have to assume that BRS and/or SUSY invariance are destroyed in the renormalized theory. This forces us and allows us to start from a more general Lagrangean having only those symmetries which are respected by the renormalization procedure (i.e. Lorentz invariance and a global $SU(2) \times SU(2)$). The anomalies automatically satisfy Wess-Zumino consistency conditions which do or do not guarantee that the parameters in the effective action can be chosen in such a way that the renormalized theory is BRS and SUSY invariant (cohomology problem).

In spite of all the complications due to the non-linearities of the SUSY transformations in the Wess-Zumino gauge it is possible to analyze this cohomology problem (although this analysis is not yet entirely completed). At present we are able to show that the anomalies for BRS transformations and field dependent gauge transformations can be removed by suitable finite renormalizations. We are confident that the same can be done for the SUSY anomaly, i.e. that there is no genuine anomaly. Once this has been achieved, all parameters of the theory are determined by symmetry requirements and by a few (gauge invariant and supersymmetric) normalization conditions.

REFERENCES

[1] O. Piguet and K. Sibold, *Nucl. Phys.* **B 247** (1984) 484, *Nucl. Phys.* **B 248** (1984) 301 and *Nucl. Phys.* **B 249** (1984) 396;
O. Piguet in this volume.

[2] J. Wess and B. Zumino, *Nucl. Phys.* **B 78** (1974) 1.

[3] K. Hepp, in *Renormalisation Theory in Statistical Mechanics and Quantum Field Theory*, C. deWitt and R. Stora eds.

[4] G. 't Hooft and M. Veltman, *Nucl. Phys.* **B 44** (1972) 189;
C.G. Bollini and J.J. Giambiagi, *Phys. Lett.* **40 B** (1972) 566;
G.M. Cicuta and E. Montaldi, *Nuovo Cimento Lett.* **4** (1972) 329.

[5] P. Breitenlohner and D. Maison, *Commun. Math. Phys.* **52** (1977) 11.

[6] W. Siegel, *Phys. Lett.* **84 B** (1979) 193.

[7] W. Siegel, *Phys. Lett.* **94 B** (1980) 37.

[8] G. Altarelli, M. Curci, G. Martinelli and S. Petrarca, *Nucl. Phys.* **B 187** (1981) 461;
N. Marcus and A. Sagnotti, Caltech Preprint CALT-68-1128 (1984).

[9] A.A. Slavnov, *Teor. Mat. Fiz.* **13** (1972) 174.

[10] R. Sénéor, in this volume.

[11] C. Becchi, A. Rouet and R. Stora, *Ann. Phys.* **98** (1976) 287.

[12] J. Schwinger, *Phys. Rev.* **82** (1951) 914, *Phys. Rev.* **91** (1953) 713.

[13] Yuk-Ming P. Lam, *Phys. Rev.* **D 8** (1973) 2943.

[14] J. Lowenstein, *Commun. Math. Phys.* **24** (1971) 1.

[15] J. Wess and B. Zumino, *Phys. Lett.* **37 B** (1971) 95.

[16] S. Ferrara and B. Zumino, *Nucl. Phys.* **B 79** (1974) 413;
A. Salam and J. Strathdee, *Nucl. Phys.* **B 80** (1974) 499;
P. Fayet, *Nucl. Phys.* **B 113** (1976) 135.

[17] P. Breitenlohner and M.F. Sohnius, *Nucl. Phys.* **B 165** (1980) 483.

RADIATIVE MASS GENERATION IN SCALE INVARIANT SYSTEMS WITH SPONTANEOUS SYMMETRY BREAKDOWN

R. Collina
Dipartimento di fisica dell' Università; Genova
Istituto Nazionale di Fisica Nucleare; Sezione di Genova

INTRODUCTION

The mass generation induced by radiative corrections present a marked phenomenological interest [1]; it has been suggested in the literature that the mechanisms may explain the nature of electron's and/or pion's mass [1,2] or a partial justification of the different mass scale which are present in the unified theories. For these reasons several mechanisms of radiative mass generation have been described in perturbative quantum field theory. The phenomenon for example arises in some H. K. gauge models [3] when the classical potential energy of the scalar fields is invariant with respect to a symmetry group which contains the gauge group as a proper subgroup, this larger symmetry being violated by the other terms of the Lagrangian. It follows that there are "accidentally" more Goldstone bosons than those implied by the symmetry of the model which are not reabsorbed by the H. K. mechanism (systems with pseudo Goldstone bosons) [1].

Radiative masses are also present in particular supersymmetric models (O'Raifertaigh model) where their appearance is triggered by the presence of an Infra Red (I.R.) anomaly [4].

Here we shall be concerned with another, relevant class of models, namely those with spontaneous symmetry breakdown of the classical scale invariance. The model where this mechanism first appeared was proposed by S. Coleman and E. J. Weinberg in 1973 [5]; Subsequently many papers discussed the subject and in particular the possibility of extending this class of models to an arbitrary order in covariant perturbation theory.

A first difficulty is due to the presence of two different scale parameters: the strength of a field vacuum expectation value and the renormalization scale, which could be mixed by the radiative corrections. A regularization independent solution of this problem is obtained by the identification of a local Ward Identity (W.I.) which, at the classical level, enforces the recursive condition that the trace anomaly be given by a scale invariant operator [6].

The second main difficulty in obtaining a complete quantum extension of theories with spontaneously broken dilatation symmetry is related to the fact that they are necessarily accompanied by an I.R. instability brought about by radiative mass generation; this instability is controlled by a suitable modification of the perturbative development [7].

In this note we shall provide a rigorous description of these models, in a covariant perturbative context, first using the W.I. instrument, and successively identifying, starting from a suitable regularization, an effective scheme of computing Feynman amplitudes [8].

We focus the attention on the main features of the problem, omitting the numerous technical details which are treated in the literature.

2-THE LOCAL SCALE INVARIANCE

To illustrate the general framework let us adopt a simple reference model built with a two component scalar field

$$\underline{\Phi} = (\Phi_1, \Phi_2) \tag{1}$$

whose classical action is invariant under the scale transformation of Φ_1 and $\Phi_2 + v$ and under the inversion $\Phi_2 \to -\Phi_1$.

One first and obvious remark is that a rigid scale transformation produces, in the shifted fields, non integrable vertices; therefore we need a local description of spontaneously broken scale invariance. Then we must consider the following local transformations

$$\begin{aligned} x^\mu &\to x^\mu - \lambda^\mu(x) \\ \delta\Phi_i(x) &= \lambda^\mu(x)\partial_\mu\Phi_i(x) \qquad i = 1, 2 \end{aligned} \tag{2a}$$

which are general coordinate transformations. The natural next step is to introduce a metric field belonging to the Einstein-Riemann representation i. e.

$$\delta g^{\mu\nu}(x) = \lambda^\rho(x)\partial_\rho g^{\mu\nu}(x) - \partial_\rho\lambda^\mu(x)g^{\rho\nu}(x) - \partial_\rho\lambda^\nu(x)g^{\rho\mu}(x). \tag{2b}$$

Unfortunately this procedure is of no help in identifying scale invariant theories. Indeed the classical terms

$$\int d^4x(-\det|g_{\mu\nu}|)^{1/2}\Phi_1{}^2, \qquad \int d^4x(-\det|g_{\mu\nu}|)^{1/2}(\Phi_2 + v)^2 \tag{3}$$

are invariant under the transformation in Eq.s(2) and in the flat limit ($g^{\mu\nu} \to \eta^{\mu\nu}$) they correspond to mass terms.

A correct description, which solves our problem, is obtained in terms of tensorial densities, with suitable Weyl weights which exclude the presence of couplings such as in Eq.(3), given by

$$\phi_1(x) = (-\det|g_{\mu\nu}|)^{1/8}\,\Phi_1(x), \tag{4a}$$

$$(\phi_2(x) + v) = (-\det|g_{\mu\nu}|)^{1/8} (\Phi_2 + v), \tag{4b}$$
$$\xi^{\mu\nu}(x) = (-\det|g_{\mu\nu}|)^{1/4} g^{\mu\nu}(x) \tag{4c}$$

which transform according to

$$\delta\phi_1(x) = \lambda^\mu(x)\partial_\mu\phi_1(x) + 1/4\ \partial_\mu \lambda^\mu(x)\phi_1(x), \tag{5a}$$
$$\delta\phi_2(x) = \lambda^\mu(x)\partial_\mu\phi_2(x) + 1/4\ \partial_\mu\lambda^\mu(x)(\phi_2(x) + v), \tag{5b}$$
$$\delta\xi^{\mu\nu}(x) = \lambda^\rho(x)\partial_\rho\xi^{\mu\nu}(x) + 1/2\ \partial_\rho \lambda^\rho(x)\xi^{\mu\nu}(x) - \partial_\rho\lambda^\mu(x)\xi^{\rho\nu}(x) - \partial_\rho\lambda^\nu(x)\xi^{\rho\mu}(x). \tag{5c}$$

The use of densities also requires the introduction of a covariant derivative

$$D_\mu\phi_i(x) = (\ \partial_\mu + \omega_\mu(x)\ )\phi_i(x) \tag{6}$$

where

$$\omega_\mu(x) = -1/8\ \partial_\mu \ln(-\det|g_{\mu\nu}|) \tag{7}$$

and

$$\delta\omega_\mu(x) = \lambda^\rho(x)\partial_\rho\omega_\mu(x) + \partial_\mu\ \lambda^\rho(x)\omega_\rho(x) - 1/4\ \partial_\mu\partial_\rho\ \lambda^\rho(x). \tag{8}$$

Observe that the choice (4c) is equivalent to

$$-\det|\xi^{\mu\nu}(x)| = 1. \tag{9}$$

The scale transformation (spontaneously broken) in Eq.(5) and (8) can be summarized in a *local functional differential operator* $W^\mu(x)$ which in the flat limit $\xi^{\mu\nu} \to \eta^{\mu\nu}$, $\omega_\mu \to 0$ is given by

$$(\ W_\mu(x))_{F\perp} = 2\partial^\rho(\delta/\delta\xi^{\mu\rho} - \eta_{\mu\rho}/4\,\delta/\delta\xi^\lambda{}_\lambda) - {}^1/_4\ \partial_\mu\partial^\rho\delta/\delta\omega^\rho + \partial_\mu\phi_i\ \delta/\delta\phi_i - {}^1/_4\ \partial_\mu[\ (\phi_i(x) + v_i)\delta/\delta\phi_i\ ], \qquad v_i = v\ \delta_{i2} \tag{10}$$

The classical theory is then completely identified by the general solution of the W.I.

$$(W_\mu(x)\Gamma^{cl.})_{F\perp} = 0 \tag{11}$$

and by the stability condition of the classical vacuum, i. e.

$$\Gamma^{cl.} = -\int d^4x\{{}^1/_2\ \xi^{\mu\nu}(x)(\partial_\mu + \omega_\mu(x))(\phi_i(x) + v_i)(\partial_\nu + \omega_\nu(x))(\phi_i(x) + v_i) + a(\phi_1(x))^4 + b^2/2\ \phi_1{}^2(x)(\phi_2(x) + v)^2\ \} \tag{12}$$

and

$$\delta/\delta\phi_i \Gamma^{cl.}|_{\phi_i=0} = 0. \tag{13}$$

3 - RADIATIVE MASS AND THE PERTURBATIVE EXPANSION

Referring to the action in Eq.(12) we see that the classical theory is defined only if the vacuum is an indifferent equilibrium state for any field translation. But the quantum fluctuations select a stable equilibrium state by introducing a linear restoring force (radiative mass term). This can intuitively be seen by looking at the level curves of the classical potential ($V = a\phi_1^4 + b^2/2\phi_1^2(\phi_2+v)^2$) of the model in fig. 1.

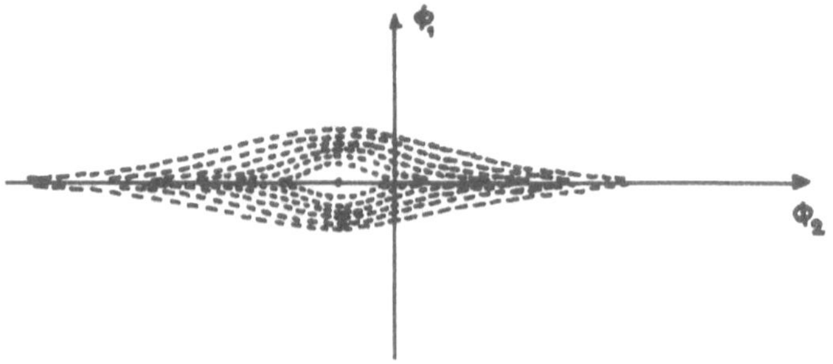

fig.1

In other words at the classical level, from Eq.(12), we have

$$m_1^2 = b^2v^2 \quad \text{and} \quad m_2^2 = 0 \; ; \tag{14}$$

but the last condition is not preserved, to the one loop level, in the ordinary perturbative expansion. Indeed at first order we have the diagrams in fig.2, where the continous and dotted lines stand for the ϕ_1 and ϕ_2 propagator respectively. Now the only counterterm containing a linear contribution in the ϕ_2 field compatible with the symmetry is

$$c(\phi_2 + v)^4 \tag{15}$$

which is necessary for the vacuum definition to the one loop level; i.e. the coefficient c

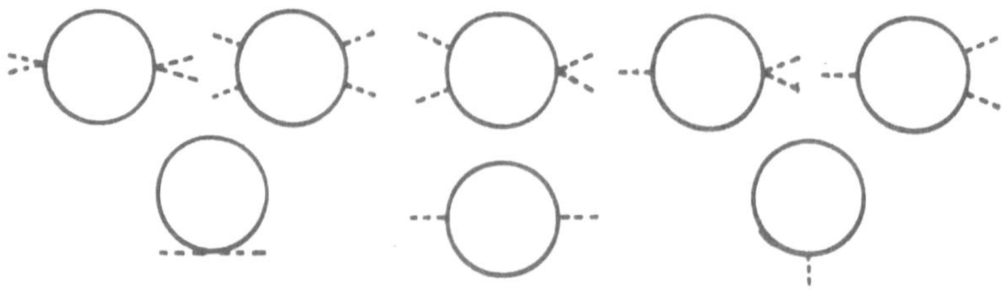

fig.2

must be chosen to cancel the contribution of the last diagram in fig.2. This choice produces finite nonvanishing contributions for the mass and vertex corrections related to the ϕ_2 field.*

Numerically we have

$$m_2^2 = (4\pi)^{-2}\hbar\, 2b^2 m^2 + O(\hbar^2)$$
$$= \hbar\, m_2^2(1) + O(\hbar^2), \qquad \text{where } m^2 = b^2 v^2 \qquad (16)$$

This radiative mass term induces an infrared sickness of the usual perturbation theory and the cure is a modified perturbative expansion [7]. Indeed the mass generation due to radiative correction can be physically interpreted only in view of a summation of the perturbative series generating a mass term in the propagator of the ϕ_2 massless particle.

The first important difference between such a perturbative approach and the usual one arises in the construction of the Fock space as a consequence of the finite mass renormalization of the ϕ_2 field, which in a formal non covariant language, could be carried out by a singular Bogoliubov transformation [9]. From a covariant point of view, this is equivalent to adding to the free lagrangian the mass term generated by the radiative corrections at lowest order; i. e. we perform the substitution in the naive massless propagator

$$p^{-2} \rightarrow (p^2 + \hbar m_2^2(1))^{-1}. \qquad (17)$$

But in this new perturbative approach the equivalence between $\hbar$ and loop ordering is not a priori guaranteed. Indeed in a completely massive model, when the propagator mass μ vanishes, a given amplitude depending from $p_1,\ldots,p_k$ external momenta, behaves like:

$$\sum_{(n\geq -n)} \mu^n\ C_{nm}\ (p_1,\ldots,p_k)(\ln\mu)^m. \qquad (18a)$$

Hence in our case the generic N-loop graph will depend on $\hbar$ as:

$$\sum_{(n\geq -n)m} \hbar^{\frac{1}{2}n+N}\, C_{nm}\, (p_1,\ldots,p_k)(\tfrac{1}{2}\ln\hbar)^m, \qquad (18b)$$

and the loop factor $\hbar^N$ may be completely hidden by $\hbar^{\frac{1}{2}n}$. For example the diagram in fig.3 has the leading power $\hbar^2$ independently of the number of loops. We have proved [7] that it is sufficient to exclude, with a suitable extra subtraction, at the one loop level the ϕ_2 propagator correction, at zero external momentum, to obtain a new consistent perturbative expansion. The resulting perturbative development is consistent in the sense that it is a formal power series in $\hbar^{\frac{1}{2}n}$ with $\ln\hbar$ corrections, and the ordering of the

* The result is scheme independent; for instance in the B.P.H.Z. renormalization, where the tadpole contributions are absent, one may fix the c parameter to maintain the vanishing of the ϕ_2 mass contribution, but then a finite trilinear ϕ_2 vertex correction survives, which is incompatible with the infrared power counting.

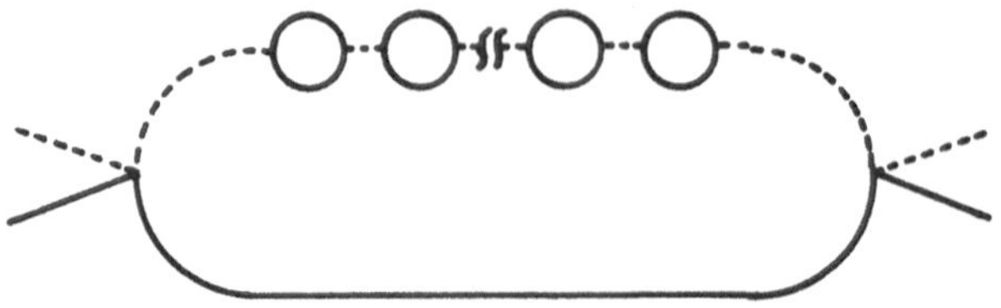

fig.3

diagrams is compatible with the loop ordering. In particular a generic proper Green function G is a formal power series in $\hbar^{\frac{1}{2}}$ of the type

$$G = G_0 + \sum_{n=1}^{\infty} \hbar^{\frac{1}{2}(n+1)} G_{(n)}(\ln\hbar) \tag{19}$$

where $G_{(n)}$ has contributions from a finite number of diagrams with at most n loops. Moreover the term $G_{(n)}(\ln\hbar)$ does not depend upon the vertices of the effective lagrangian $\mathcal{L}_{(n')}(\ln\hbar)$ for n'>n and the only contributions from $\mathcal{L}_{(n)}(\ln\hbar)$ are tree approximation terms. The proof, given in Ref.[7], is according to the B.P.H.Z. method extended to include massless particles [10].

4-RENORMALIZATION OF THE WARD IDENTITY

We return to the W.I. in Eq.(11) in order to discuss its renormalizability. The operators $W_\mu(x)$ satisfy local commutation relation i.e.

$$[\, W_\mu(x)\, ,\, W_\nu(y)\,] = -\, \partial/_{\partial x^\mu}\, \delta(x-y)\, W_\nu(x) - \partial/_{\partial x^\nu}\, \delta(x-y)\, W_\mu(y). \tag{20}$$

Let assume that the vertex functional of the renormalized theory satisfy

$$W_\mu{}^Q(\hbar,x)\Gamma = 0 \tag{21}$$

where $W_\mu{}^Q(\hbar,x)$ are a quantum extension of the operators $W_\mu(x)$, in particular

$$[W_\mu{}^Q(\hbar,x),\, W_\nu{}^Q(\hbar,y)] = -\, \partial/_{\partial x^\mu}\, \delta(x-y)\, W_\nu{}^Q(\hbar,x)\ -\ \partial/_{\partial x^\nu}\, \delta(x-y)\, W_\mu{}^Q(\hbar,y) \tag{22a}$$

and

$$\lim_{\hbar\to 0} W_\mu{}^Q(\hbar,x) = W_\mu(x). \tag{22b}$$

It is easily seen that the operators $W_\mu{}^Q(\hbar,x)$ are not unique, indeed they are not invariant under a local redefinition of the sources such as

$$J_i(x) \rightarrow J_i(x) + \epsilon\, Z(\xi(x))\, J_i(x), \tag{23a}$$
$$\xi^{\mu\nu}(x) \rightarrow \xi^{\mu\nu}(x) + \epsilon\, Z^{\mu\nu}(\xi(x)), \tag{23b}$$
$$\omega_\mu(x) \rightarrow \omega_\mu(x) + \epsilon\, Z^{\nu}{}_{\mu}(\xi(x))\omega_\nu(x) \tag{23c}$$

where $J_i(x) = - \delta/\delta\phi_i(x)\Gamma$ is the variable conjugated to $\phi_i(x)$ in the Legendre transformation. The substitutions in Eq.s(23) performed in the $Z[J]$ functional (or $\Gamma[\phi]$) amount to a new choice of the time-ordered products and have no effect on the physical interpretation of the model. The question is then, if all the possible quantized versions of the theory correspond to $W_\mu{}^Q(\hbar,x)$ operators which are related by source transformations as in Eq.s(23).

In fact we prove that the anomalies of the dilatation W.I. in Eq.(11) are related to an instability of Weyl's representation for the metric field, which can be reabsorbed by a source redefinition [6]. In particular for the anomalies we have

$$(W_\mu(x)\Gamma)_{F.L.} = \partial_\mu I(x), \tag{24}$$

where the subscript F.L. is for flat limit and $I(x)$ is a local operator obeying the minimality conditions of being of canonical dimension 4 and which cannot be written as a divergence (e.i. $I(x) \neq \partial_\mu K^\mu(x)$). $I(x)$ is the well known Callan-Symanzik or trace anomaly [11]. This anomaly can be reabsorbed by a source redefinition. Indeed adding to the Lagrangian in Eq.(12) the coupling $(\xi^{\mu\nu}(x) - \eta^{\mu\nu})I(x)$ and considering the altered representation of the metric field

$$\delta\xi^{\mu\nu}(x) = \lambda^\rho(x)\partial_\rho\xi^{\mu\nu}(x) + \tfrac{1}{2}(1+\hbar)\partial_\rho\lambda^\rho(x)\xi^{\mu\nu}(x) - \partial_\rho\lambda^\mu(x)\xi^{\rho\nu}(x) - \partial_\rho\lambda^\nu(x)\xi^{\rho\mu}(x) \tag{25}$$

we obtain, at the lowest order in $\hbar$ and in the flat limit, the new anomaly free W.I.

$$W_\mu{}^Q(\hbar,x)\Gamma\,\big|_{F.L.} \equiv \left(W_\mu(x) - \tfrac{1}{2}\partial_\mu \delta/\delta\xi\lambda_\lambda\right)\Gamma\,\big|_{F.L.} = 0. \tag{26}$$

But the description of the anomaly by the term $-\tfrac{1}{2}\partial_\mu \delta/\delta\xi\lambda_\lambda\Gamma$ does not guarantee its minimality and we are forced to analyze the W.I. outside of the flat limit. An alternative way, which is sufficient to solve the problem without going out of the flat limit, consists in the characterization of the scalar operators by introducing a source for them; i.e. a classical scalar field $\Sigma(x)$ with vanishing canonical dimension [6].

The strategy is that of trying to reabsorb the trace anomalies by transferring the instability of the metric field representation to the $\Sigma(x)$ field. Thus we put

$$\delta\Sigma(x) = \lambda^\rho(x)\partial_\rho\Sigma(x) - \hbar\partial_\rho\lambda^\rho \tag{27}$$

and analyze, in the flat limit, the new W.I.

$$(W_\mu(x) + \partial_\mu\Sigma(x)\delta/\delta\Sigma(x) + \hbar\, \partial_\mu\delta/\delta\Sigma(x))\Gamma\,|_{F\perp} = 0. \qquad (28)$$

This identity is resolved, order by order, by a Lagrangian containing arbitrary powers of Σ, but depending upon a finite number of parameters. The iterative procedure analyzed in Ref.[6] shows that the order $\hbar^N\Sigma^M$ is completely identified by the invariant term at the order $\hbar^{N+1}\Sigma^{M-1}$; the free parameters are those of the scale-invariant original theory, plus a finite number of the physically irrelevant terms containing derivative couplings for the Σ field.

5-DIMENSIONAL RENORMALIZATION

In the previous section we have illustrated the steps leading to a complete formal proof of renormalizability for a model with spontaneous symmetry breaking of scale invariance, but the problem of identifying, starting from a suitable regularization, an effective scheme of computing Feynman amplitudes remains open. To this end, and also in view of analyzing gauge models, a dimensional renormalization scheme appears as a natural choice. In this context a clear definition of scale invariance for generic space-time dimension d is required. According to the lines proposed before the behavior of a field under infinitesimal local scale transformation is specified by assigning the Weyl weight. Then a regularized version of the local scale transformation is identified by a Weyl weight depending on the space time dimension d. For example for a scalar field $\phi_i(x)$

$$\delta\phi_i(x) = \lambda^\mu(x)\partial_\mu\phi_i(x) + {}^{(d-2)}/_{2d}\; \partial_\mu\, \lambda^\mu(x)(\phi_i(x) + v_i). \qquad (29)$$

This choice assure that, in the flat limit, the operator $\partial_\mu\phi_i\partial^\mu\phi_i$ is scale invariant, but the operator $\phi_1{}^4$ has weight ${}^{(2d-4)}/_d$ instead of 1, as necessary for scale invariance. The way out of this difficulty is to introduce, already at the classical level, a spurion field $\Omega(x)$ with non vanishing vacuum expectation value and carring the Weyl weight necessary to compensate that of the operator $\phi_1{}^4$. i.e. the new invariant operator is $\phi_1{}^4\,(\Omega + \mu^{(4-d)/2})^2$ where μ is the scale parameter. It is convenient to define the dimensionless external field

$$\mu^{(4-d)/2}\Sigma(x) = \Omega(x). \qquad (30)$$

Referring to our simple model we have the new local scale transformations (with spontaneous breakdown)

$$\delta\phi_i(x) = \lambda^\mu(x)\partial_\mu\phi_i(x) + {}^{(d-2)}/_{2d}\; \partial_\mu\, \lambda^\mu(x)(\phi_i(x) + \mu^{(d-4)/2}v_i), \qquad (31a)$$

$$\delta\xi^{\mu\nu}(x) = \lambda^\rho(x)\partial_\rho\xi^{\mu\nu}(x) + {}^2/_d\; \partial_\rho\, \lambda^\rho(x)\xi^{\mu\nu}(x) - \partial_\rho\lambda^\mu(x)\xi^{\rho\nu}(x) - \partial_\rho\lambda^\nu(x)\xi^{\rho\mu}(x), \qquad (31b)$$

$$\delta\omega_\mu(x) = \lambda^\rho(x)\partial_\rho\omega_\mu(x) + \partial_\mu \lambda^\rho(x)\omega_\rho(x) - {}^{(d-2)}/_{2d}\, \partial_\mu\partial_\rho \lambda^\rho(x), \tag{31c}$$

$$\delta\Sigma(x) = \lambda^\rho(x)\partial_\rho\Sigma(x) + {}^{(4-d)}/_{2d}\, \partial_\rho \lambda^\rho(x)(\Sigma(x)+1). \tag{31d}$$

In Eq.(31a) the mass scale parameter μ has been introduced to mantain the mass dimension of v independent of the space-time dimensionality. Hence the new W.I. in the flat limit is

$$\begin{aligned}&2\partial^\rho({}^\delta/_{\delta\xi\mu\rho}\Gamma^{CL.} - \eta_{\mu\rho}/_d{}^\delta/_{\delta\xi\lambda_\lambda}\Gamma^{CL.}) - {}^{(d-2)}/_{2d}\,\partial_\mu\partial^\rho{}^\delta/_{\delta\omega\rho}\Gamma^{CL.}\\ &+ \partial_\mu\phi_i\,{}^\delta/_{\delta\phi_i}\Gamma^{CL.} - {}^{(d-2)}/_{2d}\,\partial_\mu[\,(\phi_i(x) + \mu^{(d-4)/2}v_i){}^\delta/_{\delta\phi_i}\Gamma^{CL.}\,]\\ &+ \partial_\mu\Sigma(x){}^\delta/_{\delta\Sigma(x)}\Gamma^{CL.} - {}^{(4-d)}/_{2d}\,\partial_\mu[(\Sigma(x)+1){}^\delta/_{\delta\Sigma(x)}\Gamma^{CL.}] = 0\end{aligned} \tag{32}$$

and Eq.(32) can be thought to hold for any complex value of the dimension d. Setting

$$d = 4 - 2\nu \tag{33}$$

the general solution of Eq.(32), constrained by the classical vacuum condition, appears as the obvious extension, to include the spurion field, of the classical action in Eq.(12)

$$\begin{aligned}\Gamma^{CL.} = - \int d^{4-2\nu}x\{&{}^1/_2\xi^{\mu\nu}(x)(\partial_\mu + \omega_\mu(x))(\phi_i(x) + \mu^{-\nu}v_i)(\partial_\nu + \omega_\nu(x))(\phi_i(x) + \mu^{-\nu}v_i)\\ &+a\mu^{2\nu}(1+\Sigma(x))^2(\phi_1(x))^4 + b^2/_2\,\mu^{2\nu}(1+\Sigma(x))^2\phi_1{}^2(x)(\phi_2(x) + \mu^{-\nu}v)^2\,\}\\ &+ \text{a finite number physically irrelevant terms depending on } \partial\Sigma,\end{aligned} \tag{34a}$$

and

$$^\delta/_{\delta\phi_i}\Gamma^{CL.}|_{\phi_i=0} = 0. \tag{34b}$$

The classical vacuum condition in Eq.(34b) must be maintained at the higher orders as a necessary constraint for the proper vertex functional. On the other hand the only lagrangian counterterm containing a linear term in the relevant scalar field which can be introduced without affecting the W.I. is

$$\mu^{2\nu}(1+\Sigma(x))^2(\phi_2(x) + \mu^{-\nu}v)^4. \tag{35}$$

And is the choice of this term that discriminates among the possible quantum extensions of the classical model.

Then the complete bare lagrangian obeing the W.I. in Eq.(32), is, up to a multiplicative field redefinition

$$\mathcal{L}_b = \mathcal{L}_{CL.} + \mu^{2\nu}b^2\Lambda[\nu, a({}^\mu/_v)^{2\nu}, b({}^\mu/_v)^\nu]\,(1+\Sigma(x))^2(\phi_2(x) + \mu^{-\nu}v)^4. \tag{36}$$

where $\Lambda[\nu, a({}^\mu/_v)^{2\nu}, b({}^\mu/_v)^\nu]$ is a meromorphic function of ν, computed order by order in terms of tadpole Feynman diagrams.

We observe that the proposed dimensional extension acts as a regularization only for the U.V. divergencies; the I.R. problems, related to the radiative mass generation, must be taken care of, already at the bare level. According to the philosophy discussed in Section 3 the new Feynman rules include the 1-loop radiative mass $\hbar m_2^2(1)$ in the ϕ_2 propagator and the extra subtraction concerning the one loop mass correction. It is in terms of these new rules that the Λ coefficient must be evaluated. For example the diagram in fig.4 has a contribution $\hbar^3 \ln\hbar$.

fig.4

Finally let us remark that the meaning of the theory is, a priori, not evident. Indeed the W.I. in Eq.(32) is altered by any renormalization procedure which, order by order, removes at least the ν poles. It is then relevant to discuss the class of W.I. which are obtained from the regularized one through a multiplicative renormalization procedure. In Ref.[8] it is shown that the theory can be made finite with a multiplicative Σ-dependent renormalization of the fields

$$\phi_i(x) + \mu^{-\nu} v_i \rightarrow G_i(\Sigma)(\phi_i(x) + \mu^{-\nu} v_i), \tag{37a}$$

$$h^{\mu\nu}(x) \rightarrow H(\Sigma) h^{\mu\nu}(x), \quad (\text{where } h^{\mu\nu}(x) = \xi^{\mu\nu}(x) - \eta^{\mu\nu}), \tag{37b}$$

$$\omega_\mu(x) \rightarrow L(\Sigma)\omega_\mu(x), \tag{37c}$$

$$\Sigma \rightarrow M(\Sigma) = \Sigma f(\Sigma), \qquad \text{with } f(0) \neq 0 \tag{37d}$$

which produces a renormalized W.I. equivalent, up to a finite multiplicative Σ-dependent renormalization, to that discussed in Sect.4 in a regularization independent scheme. Thus we have a precise regularization procedure to analyze the phenomenon of radiative mass generation. If we are not interested in Green's functions with Σ external legs, we can set $\Sigma = 0$ and the bare lagrangian, which in the flat limit becomes

$$\mathcal{L}_b = {}^1/_2(\partial_\mu\phi_1(x)\partial^\mu\phi_1(x)) + a\mu^{2\nu}(\phi_1(x))^4 + {b^2}/_2\, \mu^{2\nu}\phi_1{}^2(x)(\phi_2(x) + \mu^{-\nu}v)^2$$
$$+ \mu^{2\nu} b^2 \Lambda[\nu, a(\mu/_v)^{2\nu}, b(\mu/_v)^{\nu}](\phi_2(x) + \mu^{-\nu}v)^4, \tag{38}$$

is the only information needed for computations.

We observe also that the main interest of utilizing a dimensional approach is in the minimal subtraction scheme. In particular P. Breitenlohner and D. Maison have shown that, in a completely massive theory which has no I.R. problems, when the minimal subtraction scheme corresponds to a multiplicative renormalization process, the related constants are mass independent[12].

In our scheme this minimality condition is apparently violated for two main reasons i.e. the presence of tadpoles, whose compensation is required already in the bare theory, and the radiative mass which requires a new propagator accompanied by an (I.R.) extra subtraction for the one loop mass correction.

Concerning the tadpole subtraction it can be automatically implemented by a suitable alteration of the Feynman rules, which consists in neglecting the tadpole contributions to Feynman graphs. For example in fig.5 are represented the one loop and two loops mass corrections of ϕ_1 and ϕ_2 fields respectively.

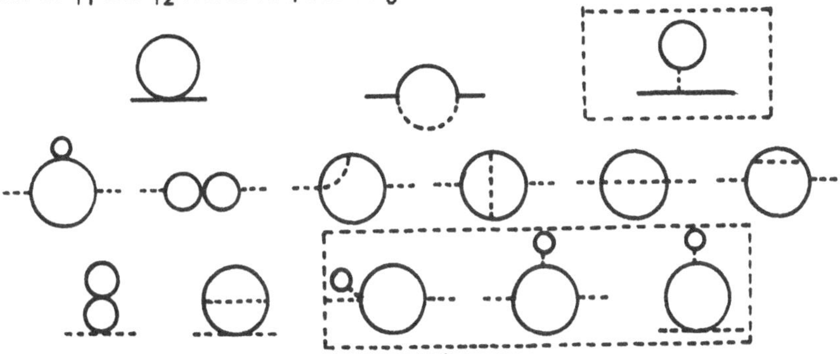

fig.5

The new propagator and the extra subtraction required by the presence of radiative mass induce, from three loops on, in the divergent part of the diagrams, terms proportional to the $\ln m^2$. But the logarithmic terms vanish in the sum of all the diagrams contributing to a given Green's function at a fixed perturbative order. The cancellation mechanism is the same as suggested by P. Breitenlohner and D. Maison for the I.R. subtraction[13]. For example in fig.6 the diagram (a) contains a divergent contribution proportional to $\hbar^3\ln(m^2/\mu^2)$ which is cancelled

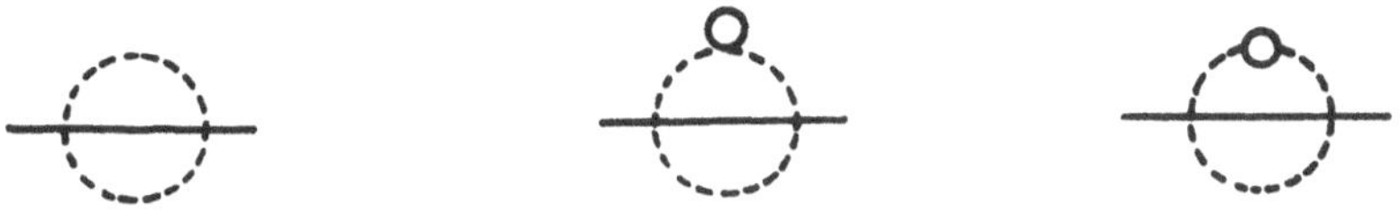

(a) (b) (c)

fig.6

by analogous terms from diagrams (b) and (c). Thus the considered renormalization procedure is minimal and the corresponding constants are also mass independent.

6-CONCLUSIONS

In this note we have analyzed in a simple model the mechanism of radiative mass generation brought about by the spontaneous breaking of dilatation symmetry in a perturbative context. The main points are the use of a local W.I. which formalizes, by means of a spurion field, the minimality requirement of the trace anomaly which appears as an instability of the classical representation of the general coordinate transformation in Weyl scheme. The consistent inclusion of the radiative mass in the propagator of the related field requires a new perturbative expansion which is compatible, although not coinciding, with the loop ordering. The modified perturbation expansion resolves the I.R. problems connected with the radiative mass.

We have also discussed how an extension to arbitrary space time dimensions d of the local dilatation W.I. provides an effective scheme for computations, which yields a quantum extension of the theory equivalent to that obtainable in a regularization independent way. An attractive feature of the procedure is that it maintains the renormalization constants mass independent.

Finally we observe that the scheme here illustrated on a simple model with only scalar fields can be directly employed to characterize to all orders the quantum extension of gauge models with spontaneously broken dilatation invariance, since the B.R.S. and W.I. operators commute. One interesting case in this class is the model proposed by S. Coleman and E. Weinberg in 1973.

Acknowledgement. I'am indebted to A. Blasi for a critical revision of the manuscript.

REFERENCES

[1] S. Weinberg - Phys. Rev. Lett.**29**(1972),1698; Phys. Rev.D**13**(1976),974; Phys. Rev. D**7**(1973),2887.
[2] H. Georgi, S. Glashow - Phys. Rev.D**7**(1973),2457.
[3] P.W. Higgs - Phys. Lett.**12**(1964),132.
T. Kibble - Phys. Rev.**155**(1967),1544.
[4] T.E. Clark, O. Piguet, K. Sibold - Nucl. Phys.B**99**(1977),292.
W.A. Bardeen, O. Piguet, K. Sibold - Phys. Lett.**72B**(1977),231.
[5] S. Coleman, J.E. Weinberg - Phys. Rev.D**7**(1973),1888.
[6] G. Bandelloni, C. Becchi, A. Blasi, R. Collina - Nucl. Phys.B**197**(1982),347.
[7] G. Bandelloni, C. Becchi, A. Blasi, R. Collina - Commun. Math. Phys.**67**(1978),147.
[8] C. Becchi, A. Blasi, R. Collina - Nucl. Phys.B**274**(1986),121.
[9] N.N. Bogoliubov - Exp. Theor. Phys. (USSR)**34**(1958),58; Nuovo Cimento**X**(1958),794.
[10] J.H. Lowenstein, W. Zimmerman - Commun. Math. Phys.**44**(1975),73.
J.H. Lowenstein - Commun. Math. Phys.**47**(1976),53.
[11] C.G. Callan, J.S. Coleman, R. Jackiw - Ann. of Phys.**59**(1970),42.
C.G. Callan - Phys. Rev.D**2**(1970),1541.
K. Symanzik - Commun. Math. Phys.**18**(1970),227.
[12] P. Breitenlohner, D. Maison - Commun. Math. Phys.**52**(1977),11.
G. Bonneau - Nucl. Phys.B**167**(1980),261; B**171**(1980),447.
[13] P. Breitenlohner, D. Maison - Commun. Math. Phys.**52**(1977),55.

Discussion session on part I:
Non-linear field transformations in 4 dimensions

To Seminar of Olivier Piguet:

In the $N = 1$ SUSY case with superfields there is an off-shell infrared problem. Writing e.g. $\mathcal{S}(\Gamma) \sim 0$ can then not be understood in the sense of scaling. I.e. the insertion Δ'_m in

$$\mathcal{S}(\Gamma) = \Delta'_m \cdot \Gamma$$

does not necessarily die out for large Euclidean (non-exeptional) momenta, because that limit does not exist.

Question: What is the precise meaning of "~ 0"?

Answer: In a context where the scaling limit cannot be performed it just means that Δ'_m has UV-power counting degree 3 instead of 4. But, in an IR-regularized theory where the scaling limit can be performed it is indeed distinguished from a hard insertion by being soft. The real goal in pure SYM would be the construction of gauge independent operators where one of the gauge parameters is the infrared regulator μ^2. There the supercurrent is of primary interest. In general SYM one would in addition expect that gauge independent matter mass insertions Δ^*_m exist which have power countig degree 3 *and* permit the scaling limit i.e. are truly soft.

To Seminar of Peter Breitenlohner:

There is an alternative approach to using directly the infinitely many non-linear field transformations and sources, namely to employ a differential algebra. This formulation has an infrared problem due to the presence of a constant anti-commuting parameter of positive dimension: it leads to the insertion of superrenormalizable vertices.

Question: Does this method really avoid the need for defining all the mentioned non-linear transformations or does it only produce the illusion that it does?

Answer: Like in the case of gauge transformation – versus BRS-transformations it really makes superfluous to define those composite operators.

Question: Where is the information referring to the SUSY content?

Answer: It is contained in the respective Slavnov identity.

Question: Should one take for serious the infrared problem and solve it?

Answer: Yes, these parameters are decisive ingredients of the complete theory. Suggestion: Make these parameters x-dependent. E.g. try the structure of *external* supergravity and perform the adiabatic limit (constant external fields). The study of the infrared problems is then at the same time a useful preparation for supergravity itself.

Part II

Non-linear σ-Models

Non-linear σ-models considered as classical field theories have a geometrical structure. The Lagrangian describes harmonic maps from space-time to a Riemannian target space with a prescribed metric. The main problem of general quantized non-linear σ-models is their non-renormalizability. In more than two space-time dimensions they are non-renormalizable by power counting. Even in two dimensions renormalization requires infinitely many parameters describing not only arbitrary changes of the coordinates (fields) but also arbitrary deformations of the metric. In order to specify a particular σ-model one has to characterize its metric within the general class of all metrics. One such possibility is provided by spaces like spheres which can be characterized by their isometry group.

THE NON-LINEAR SIGMA MODEL.

C. Becchi.
Dipartimento di Fisica - Universita di Genova and
Istituto Nazionale di Fisica Nucleare, Sezione di Genova, Italy.

Introduction

The non-linear sigma models have been introduced more than 15 years ago [1,2] to describe the infrared properties in d>2 space-time dimensions of systems with symmetry spontaneously broken according to the Golstone-Nambu mechanism.

The first step in their construction consists in the choice of a non-linear representation [2] of the spontaneously broken symmetry group. This leads to the study of models based on coset (homogeneous) spaces [3]. That is, the field carrying the Goldstone degrees of freedom belongs to the quotient space of the broken symmetry group G with respect to the stability group H of the classical vacuum configuration.

An important example of this kind is that of the chiral models in 4 space-time dimensions, where the group G is identified with the chiral e. g. SU(3)*SU(3) group and the stability group H is the diagonal (vector) SU(3). Another interestig example is the Heisenberg model in d<3 space dimensions [4]. Here G=O(n), n being the number of components of the unit spin vectors, while H=O(n-1).

More recently non linear sigma models have been discussed in d=2 dimensions to understand the physical space-time structure of string theories [5]. In this case coset spaces do not play any particular role; rather people study models where the field belongs to more general Riemannian manifolds.

From the point of view of perturbation theory, space-time dimension 2 is particularly relevant, since for d=2 the non-linear sigma models are power-counting renormalizable. As it is well known, this means that, developing the lagrangian density in powers of the field, one does not find any coefficient with negative mass dimension; not even any with positive dimension except for the infrared regulator term whose role will be discussed in the following.

Some particular models based on coset spaces, as e. g. the Heisenberg model, in 2 dimensions have been proved to be "really" renormalizable (6,7,8). By this we mean that the perturbative expansions of Green (correlation) functions of physically meaningfull operators, those independent of the particular coordinates chosen to identify the field configurations, are uniquely identified in terms of a finite number of symmetry and normalization conditions and hence depend on a finite number of parameters.

It turns also out at d=2 that, if the manifold M has "negative" curvature (that of the sphere!), the corresponding model is asymptotically free. In this situation Wilson renormalization group analysis (9) gives some interesting suggestions. First (10), above 2 dimensions, the ultraviolet properties of the model are determined by a non-trivial, ultraviolet stable fixed point and beyond perturbation theory the model seem to be renormalizable, perhaps up to 4 dimensions (4). Secondly,at d=2, the long distance properties of the theory are expected to be fixed by the possible presence of an infrared unstable fixed point. The nature of this instability has been studied (4,6,11,12) in the case of the Heisenberg model in the limit where the number of field components tend to infinity. The most relevant suggestion emerging from these studies is the presence of a dimensional transmutation mechanism generating a mass gap.

Of course the renormalization group results and suggestions refer to very particular models on coset spaces. It is by no means obvious that they could be extended to a generic compact manifold. Indeed Wilson's analysis (9) is based on the hypothesis that the relevant theory (that corresponding to a fixed point) contain only a finite number of parameters, this is not the case of a Riemannian manifold (3,13) whose geometrical properties are characterized by a generic metric tensor. Therefore one should expect to have meaningful quantum theories only for some special class of manifolds which remains to be discovered.

This problem is analogous, and perhaps strictly connected, to that of extending the class of models which are proven to be "really" renormalizable in the sense discussed above. It looks reasonable that this class would contain at least all the models based on coset spaces, since the action of these models is identified, up to a field redefinition corresponding to a coordinate transformation, by the invariance of the action under the isometries of the manifold. To the infinitesimal generators of the isometry group there correspond at the quantum level Ward identities [6] constaining the Green functions of the theory. The stability of these identities under quantum corrections guarantee the "real" renormalizability of the coset space models.

Taking as a guide the renormalization group point of view we should not be satisfied with the coset spaces. Indeed, if we assume that the renormalization group action be smooth enough to deform the field manifold without violating its global topological properties, we expect that renormalization group fixed points should exist not corresponding to coset spaces. This would happen e. g. in the case of two dimensional manifolds of genus larger than one. Referring to this situation one should wonder if the sigma models on complex algebraic curves are "really" renormalizable.

Let us also mention that a possible way to build new "really" renormalizable models could be based on the "reduction mechanism" proposed by R. Oehme and W. Zimmermann [14] and successfully applied to a vast class of theories [15].

We have given a typical example of the questions which remain open even after remarkable Friedan's thesis [16] on the renormalization of non-linear sigma models. In this thesis Friedan gives a complete set of rules, based on dimensional regularization, to characterize the possible divergences appearing in the perturbative construction of the theory. The analysis starts from the choice of a suitable coordinate system, the geodesic normal coordinates [13,17] corresponding to the bare metric of the model.

In the rest of this paper we shall recall and discuss the general aspects of the analysis of renormalization of non-linear sigma models, evidencing the results which are independent of the field parametrization. Our aim is not to present new results, but to exhibit the status of the problem in its simplest possible form.

We shall begin our analysis recalling the main formal steps of the construction of a quantum theory and the difficulties of the perturbative approach connected with the presence of divergences. Following Friedan, we shall discuss the constraints connected with the geometrical properties of the models. We shall then analyse the ultraviolet divergences appearing order by order in the perturbative expansion, comparing the case of a generic Riemannian manifold with that of a coset space.

Formal construction of the quantized theory.

The classical field is a function on R^d taking value on a C^∞ Riemannian manifold M with metric tensor g_{ij}. The classical action is defined:

$$A \equiv \frac{1}{2}\int dx^d \, g_{ij}(\phi)\, \partial_\mu \phi^i \, \partial_\mu \phi^j \tag{1}$$

A lattice regularized version (A_L) of the action is built [18] replacing R^d with the lattice Z^d. Labelling by p the points of the lattice and by μ the links, we write:

$$A_L \equiv \sum_{p,\mu} \frac{1}{2} D^2(\phi_{p+\mu}, \phi_p) \tag{2}$$

where $D(\phi, \phi')$ is the distance between the corresponding points on the manifold.

Formally the quantization of the model is based on the measure:

$$\prod_x d_g \phi(x) \; e^{-\frac{A}{\hbar}}$$

or in its lattice regularized version: (3)

$$\prod_p d_g \phi_p \; e^{-\frac{A}{\hbar}}$$

where $d_g\phi$ is the covariantly constant measure [13] on the manifold ($d_g\phi = \sqrt{\det g}\, dx^d$).

For the models of relevant interest in statistical mechanics the loop ordering parameter $\hbar$ is replaced by the "bare" temperature t [4] (which has to be renormalized as any other physical quantity).

Of course the quantum measure has to be written explicitly in terms of coordinates on the manifold ; in general this requires more than a single coordinate system which is limited to a local chart not covering M.

To avoid this difficulty we assume that the quantum fluctuations at the point x be damped when the point goes to infinity. That is: $\lim_{x\to\infty} \phi(x) = m$, where m is any point on the manifold M. We also assume that the quantum fluctuations never exceed the border of the coordinate chart centered in m. This hypothesis looks perfectly reasonable for space-time dimension $d>2$. If $d=2$, large fluctuations at infinity could be responsible of the expected infrared instabilities. In any case an assumption of this kind seems to be technically unavoidable [16]. It implies that the quantum measure is decomposed in disjoint contributions corresponding to the different asymptotic values m of the field.

We call m the "constant background field" not to be confused with ordinary variable "background field" used as a valid technical tool for calculation purposes [1,17,19].

For every constant background value m we choose a coordinate system $E_m(\sigma)$ mapping a neighborhood of the origin of R^n, if n is the dimension of the manifold, into a neighborhood of m.

In this coordinate system the metric tensor is written:

$$g_{ij}(m,\sigma) \tag{4}$$

and hence the quantum measure is:

$$d\mu_Q(m,\sigma) \equiv \prod_x d_{g_m}\sigma(x)\, e^{-\frac{1}{2\hbar}\int d^dx\, g_{ij}(m,\sigma)\,\partial_\mu\sigma^i\partial_\mu\sigma^j} \tag{5}$$

Notice that in general it is not possible to assign a coordinate system centered at every point of a manifold with continuous transition functions between every pair of overlapping systems. However this is not an obstacle to our contruction if the quantum measure is invariant under coordinate transformations. In the following we shall assume only locally the smoothness of the transition functions.

We have thus introduced in the metric tensor, and hence in the action and in the quantum functional measure, the double dependence on the field σ and on the constant background m. In other words we have independently assigned the metric tensor in every chart. It remains to assure that the different local assignements of the metric tensor correspond to a unique,globally defined tensor on the manifold M.

We find in Friedan's thesis [16] how this condition can be written in terms of a "non-linear connection" Q. Given at the point m a tangent vector v to M, Q defines a corresponding derivative acting on functions of the double variable m and σ :

$$(v,D) = (v,\partial_m) - (v,Q^i(m,\sigma))\,\partial_{\sigma^i} \tag{6}$$

Here (v, ∂_m) means the ordinary partial derivative with respect to the background m induced by the vector v, while the secon term defines a partial derivative with respect to the quantum field

The requirement that the function $f(m, \sigma)$ identifies a unique, ,globally defined, scalar function on M, is written:

$$(v, D) f(m, \sigma) = 0 \tag{7}$$

This is called the "compatibility condition" for the coordinate choice $E_m(\sigma)$.

If we assign for every point m in an open set U a basis $v^i(m)$ $(i=1,...,n)$ of $T_m(M)$, the tangent space to the manifold at the point m, we have correspondingly a system of derivatives $D^i = (v^i, D)$, and if the Lie product rules are given:

$$[v^i(m), v^j(m)] = F^{ij}_k(m) V^k(m) \tag{8}$$

we have the integrability conditions (commutation rules):

$$[D^i, D^j] = F^{ij}_k D^k \tag{9}$$

It is apparent that the compatibility condition leads directly to a functional constraint for the action and for the quantum measure.

In terms of the quantum measure one defines the correlation functions of some physically meaningful e.g. scalar function $h(m, \sigma)$:

$$G(x,y) = \int d\mu_g(m)\, d\mu_Q(m,\sigma)\, h(m,\sigma(x))\, h(m,\sigma(y)) \Big/ \int d\mu_g(m)\, d\mu_Q(m,\sigma) \tag{10}$$

where we have also integrated on the constant background m.

In the perturbative framework the correlation functions are computed in terms of Feynman amplitudes which coincide with the correlation (Green) function of the field variables. Their functional generator is the Fourier transformed quantum measure:

$$Z[m,j] = \int d\mu_Q(m,\sigma)\, e^{\int d^d x\, j_i(x)\, \sigma^i(x)} \tag{11}$$

This functional generator has only local meaning and it is not independent of the choice of local coordinates, nevertheless it is a necessary tool of perturbative renormalization.

The perturbative development of the functional generator Z is affected with two kind of divergences.

There are ultraviolet divergences which are intrinsically related to the definition of the quantum measure and have to be controlled by a suitable regularization and cured by the renormalization procedure.

We shall discuss the renormalization in the following always understanding dimensional regularization. This is indeed particularly suitable to preserve the constraints defining the quantized version of the model.

There are also infrared divergences, since the quantized field is massless and even the propagator is ill defined in two dimensions. To avoid this difficulty we shall introduce a mass term for the quantum field (6) preventing too large long wavelenght field fluctuations. This mass term ruins the compatibility condition for the action since it introduces an attractive force toward the background. In the following we shall forget this problem since the effects of the mass term are "soft" i. e. negligible at short wavelenght.

However one should remember that at the end of every computation a zero-mass limit has to be performed to recover the original geometrical structure of the theory. This limit, which has been only studied in the case of coset space models (18,20), in general does not exist, and, in the most favourable situation, it is meaningful only for some special class of correlation functions.

Renormalization.

To discuss the renormalization of our theory we have, first of all, to write the "compatibility" condition for the Feynman functional Z.

For every vector field v^i (m) we introduce a constant (m and x independent) Grassmann (anticommuting) variable C^i and an anticommuting source: $\gamma_i(x)$. Then we add to the action A the term:

$$A_s = \int d^d x \; \gamma_i(x) \; (v^j, Q^i(m, \sigma(x)))\, C_j \tag{12}$$

thus modifying the functional:

$$Z[m, j_i, \gamma_i, c_j] = \int d\mu_Q(m,\sigma)\, e^{-A_s + \int d^d x\, j_i(x)\, \sigma^i(x)} \tag{13}$$

For this new generator we have a "Slavnov-like compatibility condition" [8,21]:

$$S Z[m, j, \gamma, c_j] \equiv \left\{ c_i (v^i, \partial_m) + \int d^d x\, j_i(x) \frac{\delta}{\delta \gamma_i(x)} + \frac{1}{2} c_i c_j F^{ij}_k(m) \partial_{c_k} + \dots \right\} Z[m, j, \gamma_i, c_j] = 0 \tag{14}$$

The missing terms indicated by dots in the right-hand side have to be suitably chosen to make the S operator nihilpotent. This is always possible and the number of needed terms depends on the particular choice of the basis $\{v^i\}$ [22].

In order to make as plain as possible the formalism, in complete generality we shall choose locally vector fields generating independent translations; i. e. such that the structure constants F^{ij}_k and hence the further terms vanish. The reader should keep in mind that this has only the consequence of simplifying the formulae.

Our Slavnov identity, which is now reduced to the first two terms of Eq(14), is equivalent to the prescription of Eq(9) for the metric tensor and of the integrability condition in Eq(11).

As usual, the consequences of Eq(14) for the quantum extension of our theory are analysed introducing the vertex generator (effective action) of the theory, which is defined in terms of the connected functional: W= ln Z, by the Legendre transformation:

$$\Gamma = -W + \int d^d x\, j_i(x)\, \sigma^i(x) \quad ; \quad \sigma^i(x) = \frac{\delta W}{\delta j_i(x)} \tag{15}$$

The Slavnov identity is written in terms of Γ as follows:

$$c_i (v^i, \partial_m) \Gamma + \int d^d x \frac{\delta \Gamma}{\delta \sigma^i(x)} \frac{\delta \Gamma}{\delta \gamma_i(x)} = 0 \tag{16}$$

The analysis of this equation follows a, by now standard, iterative procedure. First one notices that Eq(16) is automatically satisfied by the "bare", dimensionally regularized, proper amplitudes. Remembering that Γ is a formal power series in $\hbar$ (t) with zeroth order value equal to the complete action $A_c = A + A_s$, and assuming that Eq(16) has not been broken by the renormalization procedure up to the (n-1)-th order, one gets for the n-th order singular terms $\Gamma_s^{(n)}$:

$$S_A \Gamma_s^{(n)} \equiv \left[c_i (v^i, \partial_m) + \int d^d x \left(\frac{\delta A_c}{\delta \gamma_i(x)} \frac{\delta}{\delta \sigma^i(x)} + \frac{\delta A_c}{\delta \sigma^i(x)} \frac{\delta}{\delta \gamma_i(x)} \right) \right] \Gamma_s^{(n)} = 0 \tag{17}$$

Since S is nihilpotent, this equation has the general solution:

$$\Gamma_S^{(n)} = S_A \, \Theta_H^{(n)} + \Omega^{(n)} \qquad \Theta_H^{(n)} = \int d^d x \, \gamma_i(x)\, \theta^{(n)i}(m, \sigma(x)) \tag{18}$$

The first term in the right-hand side is the first order variation of the action under the singular transformation: $\sigma^i \to \sigma^i + \theta^{(n)i}$, and hence it collects the terms which are trivially compensated by a field redefinition. Notice that the validity of our assumption on the independence of our theory of the particular choice of coordinates is confirmed by the fact that every possible field redefinition is automatically reabsorbed into our scheme.

The second term contains the non-trivial divergences, those affecting the coordinate independent properties of the metric tensor and of the non-linear connection. While the non-linear connection is uniquely identified by the metric tensor[14], this one can be freely deformed in an infinite number of independent ways. This in general means an infinite number of independent divergent contributions requiring each a different normalization condition.

In this situation the model is not "really" renormalizable.

It remains to discuss how the infinite number of independent normalization conditions can be replaced, in some special cases, by a finite number of constraints whose implementation makes the theory "really" renormalizable. Our discussion will be brief and necessarily limited to the, up to now, only known case, that of the coset space models.

We have already recalled that in a coset space the metric tensor is constrained and identified up to coordinate transformations by an isometry group G. To the infinitesimal generators of this group there corresponds a system of Killing vector fields [13], which in terms of our coordinates will be defined by the system of differential operators:

$$X^\alpha = X^{\alpha, i}(m, \sigma)\, \partial_{\sigma^i} \; ; \quad \text{with:} \quad [X^\alpha, X^\beta] = f^{\alpha\beta}{}_\gamma X^\gamma \tag{19}$$

not to be mistaken with the non-linear connection.

The "compatibility condition", prescribing the global definitness the

Killing vector fields, reduces to the equation: $[X^{\alpha}, D^{i}]=0$.

The isometry conditions associated with the killing fields X^{α} can be translated into a system of functional differential equations for the action:

$$X^{\alpha} A_c \Big|_{c=0} = \int d^d x \, X^{\alpha, i}(m, \sigma(x)) \frac{\delta}{\delta \sigma^i(x)} A_c \Big|_{c=0} = 0 \qquad (20)$$

In turn this system generates a system of Ward identities commuting with the Slavnov identity or, even better, after the introduction of a suitable new system of constant anticommuting coefficients, C_{α}, it can be inserted into the Slavnov identity simply adding a new term to the action:

$$A_{S'} = \int d^d x \, \gamma_i(x) X^{\alpha, i}(m, \sigma(x)) C_{\alpha} \qquad (21)$$

modifying the Slavnov operator:

$$S'Z \equiv SZ + \frac{1}{2} C_{\alpha} C_{\beta} f_{\gamma}^{\alpha\beta} \partial_{C_{\gamma}} Z = 0 \qquad (22)$$

Independently of the particular formal attitude, the substantial consequence of the existence of isometries at the quantum level is that the non trivial singular terms in $\Omega^{(m)}$ are, order by order, costrained by the same condition as the action(Eq(20)). This in general makes finite the number of independent, non-trivial, divergences and hence "really" renormalizable the theory.

Conclusion.

We now understand the different roles of "compatibility conditions" and isometry Ward identities. While the first conditions simply ensure that the matrix $g_{ij}(m,\sigma)$ appearing in the action is related to the globally defined metric tensor of the manifold without any particular constraint for the manifold itself, the Ward identities corresponding to the Killing vector fields identify the Riemannian manifold as a coset space depending on a finite number of parameters. It is this second step which makes "really" renormalizable the theory producing an essential reduction of the number of free parameters.

References.

[1] K. Meetz, J. Math. Phys. 10 (1969), 65.
J. Honerkamp, Nucl. Phys. B 36 (1972), 130.
G. Ecker and J. Honerkamp, Nucl. Phys. B 35 (1971), 481.

[2] C. G. Callan, S. Coleman, J. Wess and B. Zumino, Phys. Rev. 177 (1969), 2247.
S. Weinberg, Phys. Rev. 166 (1968),1568.
S. Coleman, J. Wess and B. Zumino, Phys. Rev. 177 (1968), 2239.

[3] S. Helgason, Differential Geometry and Symmetric Spaces, Academic Press, New York (1962)

[4] E. Brezin and J. Zinn-Justin, Phys. Rev. B 14 (1976), 3110.

[5] E. S. Fradkin and A. A. Tseytlin, Nucl. Phys. B 261 (1985), 1.
C. G. Callan, D. Friedan, E. J. Martinec and M. J. Perry, Nucl. Phys. B 262 (1985), 593.
C. Lovelace, Phys. Letters 135 B (1984), 75.

[6] E. Brezin, J. Zinn-Justin, Phys. Rev. Letters 36 (1976), 691.
E. Brezin, J. C. Guillou and J. Zinn-Justin, Phys. Rev. D 14 (1976), 2615.

[7] G. Bonnaeu and F. Delduc, Nucl. Phys. B 266 (1986), 536.
F. Delduc and G. Valent, Nucl. Phys. B 253 (1985), 494.
G. Valent, Nucl. Phys. B 238 (1984), 142.

[8] A. Blasi and R. Collina, Nucl. Phys. B (1987) in publication.

[9] K. G. Wilson and J. Kogut, Phys. Reports C 12 (1974), 75.

[10] A. M. Polyakov, Phys. Letters 59 B (1975), 79.

[11] G. Parisi, Phys. Letters 76 B (1978), 65 and Nucl. Phys. B 150 (1979), 163.

[12] F. David, Nucl. Phys. B 293 (1982), 433 and Nucl. Phys. B234 (1984), 237.

[13] A. Lichnerowicz, Theorie globale des connections et des groupes d'holonomie, Edizioni Cremonese, Roma (1962).

[14] R. Oehme, W. Zimmermann, Commun. Math. Phys. 97 (1985),569.

[15] o. Piguet and K. Sibold, this book.

[16] D. Friedan, Phys. Rev. Letters 45 (1980) 1057 and Ann. Phys.

(N. Y.) 163 (1985), 318.
[17] L. Alvarez-Gaume, D. Friedan and S. Mukhi, Ann. Phys. (N. Y.) 134 (1981), 85.
[18] S.Elitzur, Institute for Advanced Studies (1979) and Nucl. Phys. B 212 (1983), 536.
[19] P. S. Howe, G. Papadopoulos and K. S. Stelle, Nucl, Phys.

[20] F. David, Phys. Letters 96 B (1980), 371 and Commun. Math. Phys. 81 (1981), 149.
[21] A. A. Slavnov, T. M. O. 10 (1972), 153.
C. Becchi, A. Rouet, R. Stora, Phis. Letters 52 B (1974), 344.
[22] M. Henneaux, Phys. Reports 126 (1985) 2.

B.R.S. renormalization of O(n+1) non linear σ-model

A.Biasi
Dipartimento di Fisica dell'Universita'
I.N.F.N. sezione di Genova

Introduction

We propose a new method of analyzing the perturbative renormalizability of the O(N+1) non linear σ-model in two space -time dimensions. Historically the model is built with a field vector lying on the surface of the N+1 sphere of unit radius and the action is the free one[1]. The resolution of the unit length constraint yields the interaction and a particular parametrization of the action itself in terms of N independent fields. The original O(N+1) symmetry is no longer linearly realized but only a O(N) subgroup maintains a linear representation, while the remaining N generators are non linear. The model provides a simple example of non linear symmetry [2] which, in general, poses the problem of introducing a denumerable set of sources in the action in order to control the behavior under renormalization of the non linear transformations [3]. In a conventional approach this problem does not appear only in the parametrization corresponding to the orthogonal projection [4].

This we shall refer to as the algebraic problem. We shall show that replacing the commuting parameters of the transformations with anticommuting ones, allows the definition of a nihilpotent B.R.S. operator which can be used to control the algebraic properties of the model with a finite set of external fields [5].

Now the algebraic problem is not the only one connected with this model; in two dimensions the $1/k^2$ propagator is ill defined and we need a way of regularizing it. The commonly adopted procedure is to introduce a mass term with precise transformation properties under the group . The choice, natural in the orthogonal projection [4], is somewhat freer in other parametrizations [3]. Here we shall show that there is a preferred mass term and that the B.R.S. algebraic operator can be extended, still remaining nihilpotent, so that the mass term can be included in the invariant action.It will also turn out that this choice is the convenient one to discuss the zero mass limit of the theory [6] .

According to the above lines, we first consider the algebraic aspect and then the inclusion of the mass to arrive at the complete B.R.S. operator

The renormalizability of the model can now be analyzed by standard methods which, in the regularization independent approach we adopt, amount to a stability check for the classical action and to the computation of the first cohomology class.
We shall report only the results of the necessary algebraic computations whose details can be found in [7].

Finally let us remark that our method can be applied to solve the algebraic problem (modulo group theoretical difficulties) for non linear σ-models built on any symmetric or homogeneous space [8].

The algebra

The group structure of the symmetric O(N+1)/O(N) space is the following: denote by $W_{ij}= -W_{ji}$ the infinitesimal generators of the linearly realized O(N) subgroup and by W_i the remaining ones with commutation relations

$$[W_{ij},W_{kl}] = \delta_{jk}W_{il} + \delta_{il}W_{jk} - \delta_{ik}W_{jl} - \delta_{jl}W_{ik} \tag{2.1a}$$
$$[W_{ij},W_k] = \delta_{kj}W_i - \delta_{ki}W_j \qquad i;j;k;l=1...N \tag{2.1b}$$
$$[W_i,W_j] = - W_{ij} \tag{2.1c}$$

The fields of the carrier space are $\phi_i(x)$ i=1....N and the infinitesimal generators are realized as

$$W_{ij} =\int d^2x(\phi_i(x)^{\delta}/\delta\phi_j(x)-\phi_j(x)^{\delta}/\delta\phi_i(x)) = \int d^2x\rho_{ij,k}(x)^{\delta}/\delta\phi_k(x) \tag{2.2}$$

so that $\phi^2(x)= \phi_i(x)\ \phi_i(x)$ is invariant under W_{ij} ,and

$$W_i = \int d^2x[\lambda(\phi^2)\ \delta_{ik}+\sigma(\phi^2)^{\phi_i\phi_k}/_{\phi^2}]\ ^{\delta}/\delta\phi_k(x)= \int d^2x\rho_{ik}{}^{\delta}/\ \delta\phi_k(x) \tag{2.3}$$

The algebraic commutation relations in (1.1) impose the condition

$$2[\lambda(\phi^2) +\sigma(\phi^2)]\lambda'- \lambda(\phi^2)\ \sigma(\phi^2)\ /\phi^2=-1 \tag{2.4}$$

where $\lambda'={}^{d\lambda}/_{d\phi^2}$ and there is only one arbitrary function (say λ) in the model. Any choice of λ corresponds to a particular projection of the original N+1 sphere: for example $\lambda=(1-\phi^2)^{1/2}$ or

$\lambda=1/2(1-\phi^2)$ identify the orthogonal and stereographic projections respectively. The classical action invariant under (2.2-2.3) is given by

$$A= {}^1/_g \int d^2x [g_{ij}(x)\partial_\mu\phi_i(x)\, \partial_\mu\phi_j(x)] \tag{2.5}$$

with

$$g_{ij}(x)= (\lambda^2+ \phi^2)^{-1}\left[\delta_{ij} + \left({}^{(\lambda^2+ \phi^2)}/_{(\lambda+\sigma)^2} -1\right){}^{\phi_i\phi_j}/_{\phi^2}\right] \tag{2.6}$$

In order to unify notation, introduce a single greek index α={i:ij} so that the commutators in (2.1) are written as

$$[W_\alpha,W_\beta]=f_{\alpha\beta}{}^\gamma W_\gamma \tag{2.7}$$

These relations can be embedded in a "rigid" B.R.S. scheme by means of anticommuting parameters C^α, where C^{ij} is coupled to W_{ij} and C^i to W_i and it is convenient to assign to the C^α a negative unit of Faddev-Popov charge.
We now define the B.R.S. operator

$$s = C^\alpha W_\alpha - {}^1/_2\, C^\alpha C^\beta f_{\alpha\beta}{}^\gamma\, \partial/\partial C^\gamma \tag{2.8}$$

whose nihilpotency is a direct consequence of (2.7) and the Jacobi identity for the structure constants $f_{\alpha\beta}{}^\gamma$.

The classical action in (2.5) is the general Faddev-Popov neutral, local functional invariant under s .

Due to the nonlinear kernel of W_i, the s operator cannot be directly employed to discuss the renormalizability of the model, but it has to be translated in a suitable functional form with the help of a set of auxiliary external fields $\gamma_i(x)$,i=1....N,carrying a positive unit of Faddev Popov charge and assigned a canonical dimension equal to two. The new classical action becomes

$$\Gamma_{cl}= A + C^\alpha \int d^2x \rho_{\alpha i}(x)\, \gamma_i(x) \tag{2.9}$$

and the B.R.S. symmetry is described by the functional relation

$$(S\,\Gamma_{cl})= \int d^2x (\delta\,\Gamma_{cl}/\delta\gamma_i(x)\; \delta\,\Gamma_{cl}/\delta\phi_i(x) + {}^1/_2 C^\alpha C^\beta f_{\alpha\beta}{}^\gamma \partial\Gamma_{cl}/\partial C^\gamma \tag{2.10}$$

To check the nihilpotency of the operator in (2.10) we have to rewrite it for the connected functional $Z_{cl}[J] = \Gamma_{cl} + \int d^2x J_i(x)\phi_i(x)$ evaluated at $J_i(x)= -\delta\,\Gamma_{cl}/\delta\phi_i(x)$, where it reads

$$S\, Z_{cl}[J]=\left[\int d^2x\left(-J_i(x)\,\delta/\delta\gamma_i(x)\right)+1/2\, C^{\alpha}\, C^{\beta}\, f_{\alpha\beta}{}^{\gamma}\,\partial/\partial C_{\gamma}\right]Z_{cl}[J] \tag{2.11}$$

We have now a functional identity which could be used as the basis to analyze the quantum corrections to the classical action in (2.9); as mentioned in the Introduction we still to define the propagator with the addition of an appropriate mass term to the action.

The mass term

Let us first notice that the nihilpotency of the BRS operator can be easily exploited to introduce in the action an arbitrary mass term $m^2 f(\phi^2)$ provided we also introduce its variation coupled to an external field. This is the approach considered in [7]; here we shall illustrate an alternative procedure which identifies a preferred mass term [9].

There is one projection , namely the orthogonal [4],where the natural choice is to select the mass along the projection axis; here the mass is given by $(1-\ \phi^2)^{1/2}$ and its transformations are

$$W_i(1-\phi^2)^{1/2}=-\phi_i \qquad W_i\,\phi_k=(1-\phi^2)^{1/2}\,\delta_{ik} \tag{3.1}$$

In order to reproduce (3.1) in an arbitrary parametrization, we introduce composite operators(interpolating fields)

$$\Xi_i(x)=\phi_i(x)\,G(\phi^2(x)) \tag{3.2}$$

and impose at the classical level the analog of (3.1),i.e.

$$W_i\,M(\phi^2)=-\,\Xi_i \qquad W_i\Xi_k=\,M(\phi^2)\delta_{ik} \tag{3.3}$$

The solution to (3.3) is easily found to be

$$G=M/\lambda \quad \text{and} \quad M'/M=-\,1/2\lambda(\lambda+\sigma) \tag{3.4}$$

which identifies both the mass term and the interpolating field up to a multiplicative constant.

To implement (3.4) beyond the classical level we need sources $K_i(x)$ for the composite operators $\Xi_i(x)$ and $\omega(x)$ for the mass term which are assigned a canonical dimension equal to two and are Faddev Popov neutral.
The complete classical action is now

$$\Gamma_m = \int d^2x K_i(x)\Xi_i(x) + \int d^2x \omega(x) M(\phi 2) + \Gamma_{cl} \tag{3.5}$$

with Γ_{cl} given in (2.9). The behavior of the first two terms on the r.h.s. of (3.5) under infinitesimal transformations is computed by (2.2;2.3;3.3);explicitely we have

$$W_{ij}\int d^2x K_i(x)\Xi_i(x) = \int d^2x (K_i(x)\Xi_j(x) - K_j(x)\Xi_i(x)) = \int d^2x (K_i \delta/\delta K_j - K_j \delta/\delta K_i)\Gamma_m \tag{3.6a}$$

$$W_i \int d^2x K_i(x)\Xi_i(x) = \int d^2x K_i(x) M(\phi 2) = \int d^2x \delta/\delta\omega \Gamma_m \tag{3.6b}$$

$$W_{ij}\int d^2x \omega(x) M(\phi 2) = 0 \tag{3.7a}$$

$$W_i \int d^2x \omega(x) M(\phi 2) = -\int d^2x \omega(x)\Xi_i(x) = -\int d^2x \omega(x) \delta/\delta K_i \, \Gamma_m \tag{3.7b}$$

According to the above expressions ,we modify the BRS operator in(2.11) to the new form

$$S = \int d^2x (-J_i(x)\delta/\delta\gamma_i(x)) + {}^1/_2 C^\alpha C^\beta f_{\alpha\beta}{}^\gamma \partial/\partial C_\gamma + 2C^{ij}\int d^2x K_j \delta/\delta K_i + C^i \int d^2x K_i \delta/\delta\omega$$
$$C^i \int d^2x (\omega + m^2)\delta/\delta K_i \tag{3.8}$$

where we have shifted $\omega(x)$ to $\omega(x)+m^2$ in order to have an explicit mass parameter.A direct check shows that S^2 still vanishes.

The proof of the renormalizability of the model is now based on the possibility of extending to all perturbative orders the identity

$$SZ[J_i;\gamma_i;K_j;\omega] = 0 \tag{3.9}$$

for the connected functional Z,or

$$(S\Gamma) = \int d^2x (\delta\Gamma/\delta\gamma_i(x) \delta\Gamma/\delta\phi_i(x) + {}^1/_2 C^\alpha C^\beta f_{\alpha\beta}{}^\gamma \partial\Gamma/\partial C_\gamma + 2C^{ij}\int d^2x K_j \delta\Gamma/\delta K_i$$
$$+ C^i \int d^2x K_i \delta\Gamma/\delta\omega - C^i \int d^2x (\omega + m^2)\delta\Gamma/\delta K_i \tag{3.10}$$

for the vertex functional Γ whose classical approximation is given in (3.5).

The procedure we shall follow is the one adopted when no regularization is assumed and it requires two independent checks; first we must control that the classical model is stable and then that the first cohomology class of the S operator is empty [10].

In our case the computation is completely algebraic i.e. the group structure is sufficient to guarantee the above conditions.

Stability

We shall analyze this property by means of the identity (3.10) for the vertex functional; notice that the expression for $(S\Gamma)$ is invariant under the field, source and parameter redefinition

$$\phi_i(x) \to \phi_i(x)\, Z(\phi^2(x)) \tag{4.1a}$$
$$\gamma_i(x) \to \eta_{ij}(x)\, \gamma_j(x) \quad \text{with } \eta_{ij}(x)=Z(\phi^2(x))\, \delta_{ij}+Z'(\phi^2(x))\, \phi_i(x)\, \phi_j(x) \tag{4.1b}$$
$$K_i(x) \to a\, K_i(x) \tag{4.1c}$$
$$\omega(x) \to a\, \omega(x) \tag{4.1d}$$
$$m^2 \to a\, m^2 \tag{4.1e}$$
$$g \to Z_g\, g \tag{4.1f}$$

Suppose we now perturb the classical in (3.5) with a term

$$\epsilon\Gamma_1= \int d^2x A_{ij}(x)\partial_\mu\phi_i(x)\, \partial_\mu\phi_j(x)+ C^\alpha \int d^2x R_{\alpha i}(x)\, \gamma_i(x) + \int d^2x K_i(x) B_i(x)$$
$$+ \int d^2x \omega(x) E(\phi^2)+ m^2 \int d^2x F(\phi^2) \tag{4.2}$$

which to first order in ϵ satisfies

$$(S[\Gamma_m+\epsilon\Gamma_1])=0 \tag{4.3}$$

where $A_{ij}(x)$, $R_{\alpha i}(x)$, $B_i(x)$, $E(x)$, $F(x)$ are functions of the fields $\phi_i(x)$ without space-time derivatives.

The model is said to be stable if such perturbation can be reabsorbed by performing on the classical action a transformation as in (4.1).

It is well known that stability is a necessary condition for renormalizability and that it becomes also sufficient if there is a regularization prescription which preserves the BRS identity.

To proceed with the stability check, we select in (4.3) the first order in ϵ and obtain

$$S_L\Gamma_1=\{\int d^2x(\delta\Gamma_m/\delta\gamma_i(x)\,\delta/\delta\phi_i(x))+\int d^2x(\delta\Gamma_m/\delta\phi_i(x)\,\delta/\delta\gamma_i(x))+1/2C^\alpha C^\beta f_{\alpha\beta}{}^\gamma\,\partial/\partial C_\gamma+2C^{ij}\int d^2xK_j\,\delta/\delta K_i$$
$$+C^i\int d^2xK_i\,\delta/\delta\omega - C^i\int d^2x(\omega+m^2)\,\delta/\delta K_i\}\Gamma_1 \qquad (4.4)$$

where the linearized operator S_L is still nihilpotent.

We discuss the solution of (4.4) by setting to zero the coefficients of the independent expressions in the external fields and anticommuting parameters. First select the coefficient of $\gamma_i(x)C^\alpha C^\beta$ which yields the system of equations

$$W_\beta R_{\alpha i}(x) - W_\alpha R_{\beta i}(x) - \int d^2y R_{\alpha j}(y)\,\delta/\delta\phi_j(y)\,\rho_{\beta i}(x)+ \int d^2y R_{\beta j}(y)\,\delta/\delta\phi_j(y)\,\rho_{\alpha i}(x)$$
$$+ f_{\alpha\beta}{}^\tau R_{\tau i}(x) = 0 \qquad (4.5)$$

whose solution, analyzed in detail in [7],is

$$R_{ij,k}=0 \qquad R_{ij}(x)=\Lambda(\phi^2(x))\delta_{ij} + \Sigma(\phi^2(x))\phi_i(x)\phi_j(x) \qquad (4.6a)$$

with

$$2[\Lambda'(\sigma+\lambda)+\lambda'(\Sigma+\Lambda)] - (\Sigma\lambda+\sigma\Lambda)/\phi^2 =0 \qquad (4.6b)$$

Notice that the constraint (4.6b) corresponds to the first order perturbation $\lambda \to \lambda+\epsilon\Lambda$ $\sigma \to \sigma+\epsilon\Sigma$ ofthe algebraic closure condition (2.4). These terms can be reabsorbed by the choice

$$Z(\phi^2(x))=1+b(x) \quad \text{with } (\lambda-2\lambda'\phi^2)b=\Lambda \qquad (4.7)$$

Suppose we perform the substitution (4.7) in the classical action so that the contribution $C^\alpha\int d^2xR_{\alpha i}(x)\gamma_i(x)$ disappears from Γ_1 and $Z(\phi^2(x))$ is now fixed; we then select the coefficients of $C^\alpha K_i(x)$ and $C^\alpha\omega(x)$ which yield the equations

$$W_{ij}B_l(x) - 2\delta_{il}B_j(x) =0 \qquad (4.8a)$$
$$W_jB_i(x) - \delta_{ij}E(x) =0 \qquad (4.8b)$$
$$B_i(x) + W_iE(x) =0 \qquad (4.8c)$$

with solutions:

$$B_i(x) =\phi_i(x)B(\phi^2(x)) \qquad (4.9a)$$
$$E(x) =\lambda B(\phi^2(x)) \qquad (4.9b)$$

provided

$$E'/E= -1/2\lambda(\lambda + \sigma) \qquad (4.10)$$

Comparing (4.9:4.10) with (3.4) we have $E=kM$ and $B=kM/\lambda$.

Finally the external field independent contributions to (4.4) satisfy

$$W_\alpha \int d^2x A_{ij}(x)\partial_\mu\phi_i(x)\,\partial_\mu\phi_j(x) = 0 \tag{4.11a}$$
$$\int d^2x B_i(x) + W_i \int d^2x F(x) = 0 \tag{4.11b}$$

which imply

$$F(x) = E(x) \tag{4.12a}$$
$$A_{ij}(x) = r g_{ij}(x) \tag{4.12b}$$

All these terms can be generated from the classical action through the transformations in (4.1) with $a = 1+\epsilon k$ $\;Z_g = 1+\epsilon r$.

The stability check is now completed and we shall consider the problem of the presence of anomalies in the next section.

Cohomology

In a regularization independent approach we still have to show that the BRS invariance can be maintained at all perturbative orders, i.e. that there are no anomalies.
The steps through which this control is performed are well known and are based on an inductive procedure. The vertex functional Γ is a formal power series in the loop ordering parameter $\hbar$ with the zeroth order given by the BRS invariant classical action in (3.5).
Suppose that (3.10) holds up to the order n-1, then at the next order we find

$$(S\Gamma)^{(n)} = \Delta \tag{5.1}$$

where, by the Quantum Action Principle [11], we know that Δ is a local functional of the fields with canonical dimension less than or equal to two and with a negative unit of Faddev Popov charge.

The nihilpotency of the BRS operator insures that Δ obeys the consistency condition [12]

$$S_L \Delta = 0 \tag{5.2}$$

where S_L is the linearized operator in(4.4).

The solution of (5.2) can be written as

$$\Delta = S_L \hat{\Delta} + \Delta_* \tag{5.3}$$

and Δ_* is not a BRS variation. The number of independent terms (if any) in Δ_* is the number of anomalies and coincides with the dimensionality of the first cohomology space.

Therefore we seek the general solution of (5.1); if this implies that Δ_* vanishes than we say that the breaking Δ is compensable and the needed counterterm is exactly $\hat{\Delta}$.

We parametrize Δ and $\hat{\Delta}$ as:

$$\Delta = C^{\alpha}C^{\beta}\int d^2x\,\Delta_{\alpha\beta}{}^{i}(x)\gamma_i(x) + C^{\alpha}\int d^2x\,K_i(x)D_{\alpha i}(x) + C^{\alpha}\int d^2x\,\omega(x)A_{\alpha}(x) + m^2C^{\alpha}\int d^2x\,B_{\alpha}(x) + C^{\alpha}\int d^2x\,\Delta_{\alpha}(x) \tag{5.4a}$$

$$\hat{\Delta} = C^{\alpha}\int d^2x\,\hat{\Delta}_{\alpha}{}^{i}(x)\gamma_i(x) + \int d^2x\,K_i(x)D_i(x) + \int d^2x\,\omega(x)A(x) + m^2\int d^2x\,B(x) + \int d^2x\,\Delta_0(x) \tag{5.4b}$$

where $\Delta_{\alpha}(x)$, $\Delta_0(x)$ depend upon the fields $\phi_i(x)$ and two space time derivatives, all other coefficients being functions of $\phi_i(x)$ alone.

Insert now (5.4a) into (5.2) and isolate the independent terms beginning with the coefficient of $C^{\alpha}C^{\beta}\gamma_i(x)$ which yields

$$C^{\alpha}C^{\beta}\int d^2x\,\Delta_{\alpha\beta}{}^{i}(x)\gamma_i(x) = 8C^{ij}C^{i1}\int d^2x\,T(x)\phi_1(x)\gamma_1(x) + 2C^{i}C^{j}\int d^2x\,R(x)\phi_i(x)\gamma_j(x) \tag{5.5}$$

The expression in the r.h.s. is compensable by the choice of the corresponding term in $\hat{\Delta}$ given by

$$C^{\alpha}\int d^2x\,\hat{\Delta}_{\alpha}{}^{i}(x)\gamma_i(x) = C^{ij}\int d^2x\,T(x)\phi_i(x)\gamma_1(x) + C^{i}\int d^2x\,H(x)\phi_i(x)\phi_j(x)\gamma_j(x) + C^{j}\int d^2x\,\Theta(x)\gamma_j(x) \tag{5.6a}$$

with

$$R(x) = \sigma\Theta/\phi^2 - 2\Theta'(\lambda+\sigma) - H\lambda - 2\lambda'(\Theta + H\phi^2) \tag{5.6b}$$

The apparent undetermination in (5.6b) is easily explained if we observe that the compensability relation $\Delta = S_L\hat{\Delta}$ determines $\hat{\Delta}$, once Δ is given by the consistency condition (5.2), only up to terms $S_L\hat{\varphi}$. We can use this degree of freedom to eliminate the undetermination in $\hat{\Delta}$; indeed the only candidate with the correct quantum numbers is

$$\hat{\varphi} = \int d^2x\,L(x)\phi_i(x)\gamma_i(x) \tag{5.7a}$$

with variation

$$S_L\hat{\varphi} = C^{i}\int d^2x\left[L(2\lambda'\phi^2+\lambda)\gamma_i + (2L\sigma' - 2L'(\lambda+\sigma) - L\sigma)\phi_i\phi_k\gamma_k\right] + (\gamma_k\text{independent terms}) \tag{5.7b}$$

and by a suitable choice of L we can eliminate either H or Θ in (5.6a).

Let us analyze now the remaining contributions which do not contain space time derivatives; the consistency condition (5.2) reduces them to the form

$$C^i\int d^2x(\omega\,\phi_i A)(x) + m^2 C^i\int d^2x(\phi_i B)(x) + C^i\int d^2x K_j(x)\Big[D\delta_{ij} + (A+2D'(\sigma+\lambda))\lambda^{-1}\phi_i\phi_j\Big](x) \tag{5.8}$$

which is compensable by

$$\int d^2x\omega(x)A(x) + m^2\int d^2xB(x) + \int d^2xK_i(x)\phi_i(x)D(x) \tag{5.9}$$

provided

$$D(x)+2A'(x)(\sigma+\lambda)(x) = -A(x) \tag{5.10a}$$

$$D(x)+2B'(x)(\sigma+\lambda)(x) = -B(x) \tag{5.10b}$$

$$A(x) - D(x)\,\lambda(x) = D(x) \tag{5.10c}$$

In particular, the compensability of the last term on the r.h.s. of (5.8) is insured by the choice in (5.10) through the use of the algebraic closure relation (2.4).

Finally the external fields independent part of Δ containing two space time derivatives when inserted in (5.2) becomes

$$C^i W_i \int d^2x\,[G_{ij}(x)\partial_\mu\phi_i(x)\,\partial_\mu\phi_j(x)] \tag{5.11a}$$

which is trivially compensable by

$$\Delta_0(x) = -\,G_{ij}(x)\partial_\mu\phi_i(x)\,\partial_\mu\phi_j(x) \tag{5.11b}$$

We conclude therefore that the breaking Δ is compensable and there are no anomalies.

The proof of the renormalizability of the model is now completed if renormalizability is intended in the broader sense specified in [13]; indeed the fields $\phi_i(x)$ and the external fields $\gamma_i(x)$ have renormalization "constants" which are functions of $\phi^2(x)$.

However, looking at the list in (4.1) we see that all other external fields and parameters are renormalized with true constants which are formal power series in $\hbar$. This suggests that the interesting objects to look at are the connected Green functions which have external legs of interpolating fields $\Xi_i(x)$ i.e. the ones obtained from the connected functional by deriving w.r.t. the external fields $K_i(x)$. In this case we can set to zero all the uninteresting sources and anticommuting parameters to obtain from (3.8;3.9)

$$\Big[\int d^2xK_i\delta/\delta\omega - \int d^2x(\omega+m^2)\delta/\delta K_i\Big] Z\{J_i=0;\gamma_i=0;C^\alpha=0;K_j;\omega\} \tag{5.12}$$

which can be used to characterize the mass insertions in these Green functions in a manner identical to that proposed in [6] and thus (5.12) is the starting point to analyze the zero mass limit of the theory.

References

[1] M.Gell-Mann, M.Levy -Nuovo Cimento 16 (1960) 705

[2] S.Coleman, J.Wess, B.Zumino -Phys.Rev. 177 (1969) 2239
M.Bando, T.Kuramoto, T.Maskawa, S.Uehara -Prog.Theor.Phys. 72 (1984) 313; 72 (1984) 1207

[3] G.Bonneau, F.Delduc -Nucl.Phys. B266 (1986) 536

[4] E.Brezin, J.Zinn-Justin, J.C. LeGuillou -Phys.Rev. D14 (1976) 2615

[5] C.Becchi -These proceedings

[6] F.David -C.M.P. 81 (1981) 149

[7] A.Blasi, R.Collina -"Renormalization a la BRS of the non linear σ model" to appear in Nuclear Physics B

[8] D.H.Friedan -Annals of Phys. 163 (1985) 318
G.Valent -Nucl.Phys. B238 (1984) 142
F.Delduc, G.Valent -Nucl.Phys. B253 (1985) 494

[9] C.Becchi, A.Blasi, R.Collina - in preparation

[10] C.Becchi, A.Rouet, R.Stora -Annals of Phys. 98 (1976) 287 and "Gauge field models" in Renormalization Theory edited by G.Velo, A.S.Wightman-Reidel Publ.Co. 1976

[11] Y.M.P.Lam -Phys.Rev. D6 (1972) 2145 ; D7 (1973) 2943
J.Lowenstein -C.M.P. 24 (1971) 1

[12] O.Piguet, A.Rouet -Phys.Reports 76C (1981) 1

[13] G.Bonneau -Nucl.Phys. B221 (1983) 178

RENORMALIZATION OF BOSONIC NON-LINEAR σ-MODELS BUILT ON COMPACT HOMOGENEOUS MANIFOLDS

Guy BONNEAU

Laboratoire de Physique Théorique et Hautes Energies
Université Paris VII, Tour 24, 2 place Jussieu, 75251 PARIS CEDEX 05, FRANCE

Abstract. We review the quantum status of the non-linear bosonic σ-models built on compact homogeneous spaces. The subclass of Kähler manifolds can be parametrized in such a way that multiplicative renormalizability holds, to all-order of perturbation theory. The essential ingredients are the homogeneity of the space and the existence of a charge Y that separates the fields in ϕ and $\bar{\phi}$: for these Kähler manifolds, a family of coordinate frames exists such that the non-linear isometries are holomorphic. The method is exemplified on the special case SU(3)/(U(1)xU(1)).

The material presented here results from work done in Paris with François DELDUC and Galliano VALENT

I. INTRODUCTION AND GENERAL SURVEY

Non-linear σ-models on homogeneous (coset) spaces G/H play an essential role in the analysis of symmetry breaking : whenever the symmetries of a model are broken down from a compact group G to a closed subgroup H, the associated Goldstone bosons are described by a non-linear σ-model on the coset G/H [1]. In a 2-dimensional space time, the renormalization program can be undertaken : the aim of this talk is to enumerate what is known on that subject and give some new results, depending on the geometry of the manifold G/H. The n fields ϕ^i's being understood as coordinates on the n-dimensional real Riemannian manifold G/H whose metric is $g_{ij}[\phi]$, the invariant action is written as

$$I[\phi] = \frac{1}{2}\int d^2x\, g_{ij}[\phi]\, \partial_\mu\phi^i\, \partial_\mu\phi^j \qquad (1)$$

If $\mathcal{G}$ (resp. $\mathcal{H}$) is the Lie algebra of G (resp. H) and its generators are separated in $H_i \in \mathcal{H}$ and $X_a \in \mathcal{G} - \mathcal{H}$, the commutation relations are

$$[H_i, H_j] = f_{ij}{}^k\, H_k \qquad (2.a)$$

$$[H_i, X_a] = f_{ia}{}^b\, X_b \qquad (2.b)$$

$$[X_a, X_b] = f_{ab}{}^c\, X_c + f_{ab}{}^i\, H_i \qquad (2.c)$$

In the symmetric space case, the $f_{ab}{}^c$ vanish. Symmetric spaces are listed in [2] and the corresponding σ-models are known to be integrable [3]. These spaces are Einstein :
$R_{ij} = c\, g_{ij}$ with $c > 0$ in the compact case, which is responsible for (one-loop) asymptotic freedom. There is no general proof of renormalizability for these models in the real case

but for $S^N = SO(N+1)/SO(N)$ [4], the Grassmannian $SO(N)/(SO(p)\times SO(N-p))$ [5] and $SU(N)/SO(N)$, $SU(2N)/Sp(N)$ recently analysed [6]. Moreover, these proofs use a definite choice of parametrization of the coset space (plus dimensional regularization, that is licit for non-linear bosonic σ-models). A promising approach *à la B.R.S.* allowed A. Blasi and R. Collina to give a renormalization proof for S^N case that does not rely upon any regularization and is independent of the choice of a parametrization of the sphere - as soon as the sugroup H is linearly realized [7].

On the contrary, in the hermitian(*) symmetric case, a complete analysis and renormalizability proof was given by F. Delduc and G. Valent [9] using a special parametrization of the coset space where the isometries are holomorphic in ϕ^α and $\bar{\phi}^{\bar{\alpha}}$ (the complex coordinates adapted to the hermitian structure that exists in that case [8]). Moreover, these hermitian symmetric spaces are known to be Kähler(*) manifolds (ref.[2] page 372).

In our way towards an unified treatment of non-linear σ-models built on a coset space G/H, we first looked at homogeneous Kähler manifolds. For a systematic study of complex homogeneous spaces, the following mathematical results are in order :

Theorem I

Let G be a compact connected semi-simple Lie group and H a proper closed connected subgroup. Then three equivalent propositions are :

(i) G/H is a complex homogeneous space and G and H have the same rank,

(ii) G/H is Kähler,

(iii) H is the centralizer of a torus(**) in G.

The equivalence (i) ⇔ (iii) is proved in ref.[10.a], the equivalence (iii) ⇔(ii) in ref.[10.b]. Notice that hermitian symmetric spaces [9] correspond to a one-dimensional torus [11].

Homogeneous Kähler spaces are widely discussed in the mathematical litterature but have been studied only recently in theoretical physics for supersymmetric model building [11]. Here we want to emphasize the pioneering work of M. Bando, T. Kuramoto, T. Maskawa and S. Uehara [12], hereafter referred to as B.K.M.U. The classical analysis of bosonic homogeneous Kähler σ-models relies on their method which was also cleared up in K. Itoh, T. Kugo and H. Kunitomo's work [13]. Indeed, the B.K.M.U. method offers a parametrization of the manifold adapted to the hermitian structure and where, as in the symmetric case [9], the isometries are holomorphic.

In the next section, we shall exemplify the general method with the $SU(3)/(U(1)\times U(1))$ case (which is Kähler, see theorem I). Using B.K.M.U. parametrization of a coset space, we obtain the classical action depending on 3 parameters and the expression of the isometries - which are of the desired holomorphic form : $\delta\phi = f(\phi)$, $\delta\bar{\phi} = \bar{f}(\bar{\phi})$. A detailed analysis of this model will be presented elsewhere [14].

(*) An introduction to complex manifolds and Kähler geometry can be found in these proceedings ([8] and references therein).

(**) A torus means a direct product of any U(1) subgroups of G, $(U(1))^k$ with $k \leq r$= rank G. Its centralizer means the subgroup of all the elements of G that commute with these U(1)'s. It has clearly the same rank as the group G.

In section III, we then explain the B.K.M.U. construction of the general homogeneous Kähler σ-model. The isometries are holomorphic :

$$\delta\phi^{\alpha} = \varepsilon^{\alpha} + F^{\alpha}_{\beta\gamma}(\phi)\,\varepsilon^{\beta}\phi^{\gamma} + G^{\alpha}_{\bar\beta\gamma}(\phi)\,\bar\varepsilon^{\beta}\phi^{\gamma} \tag{3}$$

where $F^{\alpha}_{\beta\gamma}(\phi)$ and $G^{\alpha}_{\bar\beta\gamma}(\phi)$ are finite order polynomials in ϕ, and ε and $\bar\varepsilon$ are the parameters of the non-linear transformations. In the hermitian symmetric case, $F^{\alpha}_{\beta\gamma}(\phi)$ vanishes and $G^{\alpha}_{\bar\beta\gamma}(\phi)$ is linear in ϕ : $G^{\alpha}_{\bar\beta\gamma}(\phi) = \frac{1}{2}R^{\alpha}{}_{\gamma\delta\bar\beta}\,\phi^{\delta}$ where $R^{\alpha}{}_{\gamma\delta\bar\beta}$ is the Riemann curvature of the manifold at $\phi = \bar\phi = o$ [9]. We also explain the origin of the Kähler potential.

In section IV, using dimensional regularization and the holomorphic expression (3) of the isometries in B.K.M.U. parametrization (also valid in any holomorphically related parametrization that keeps the linearly realized H transformations), we are able to prove the <u>all-order multiplicative renormalizability, with no field renormalization at all.</u> This new result comes essentially from two facts :

(i) the existence of a charge that separates ϕ and $\bar\phi$, that is to say the complex character of the manifold,

(ii) the existence of a (class of) coordinate system(s) in which the non-linear isometries are holomorphic, a property that results from the Kähler character of the manifold.

II. CLASSICAL ANALYSIS OF THE HOMOGENEOUS SPACE SU(3)/(U(1)xU(1))

This six-dimensional real manifold is a complex three-dimensional Kähler one (theorem I). It is then natural to choose complex coordinates $\phi^{\alpha},\bar\phi^{\bar\alpha}$ adapted to the hermitian structure [8] : in such coordinates, if the isometries are holomorphic, the hermitian structure of the metric will be stable.

The su(3) algebra is the commutator algebra of 3 x 3 antihermitian traceless matrices. We take for $\mathcal{H}$ the subalgebra of diagonal matrices spanned by iH_1 and iH_2

$$H_1 = \begin{pmatrix} 1 & 0 & 0 \\ 0 & -1 & 0 \\ 0 & 0 & 0 \end{pmatrix} \qquad H_2 = \begin{pmatrix} 1 & 0 & 0 \\ 0 & 1 & 0 \\ 0 & 0 & -2 \end{pmatrix} \tag{4}$$

and choose (*) to parametrize the orthogonal complement $(\mathcal{G} - \mathcal{H})_{\perp}$ of $\mathcal{H}$ in $\mathcal{G}$ by :

$$\begin{pmatrix} 0 & -\bar\varepsilon^{3} & -\bar\varepsilon^{1} \\ \varepsilon^{3} & 0 & -\bar\varepsilon^{2} \\ \varepsilon^{1} & \varepsilon^{2} & 0 \end{pmatrix} = \varepsilon^{\alpha}X_{\alpha} + \bar\varepsilon^{\alpha}X_{\bar\alpha}\ , \quad X_{\bar\alpha} = -(X_{\alpha})^{\dagger} \tag{5}$$

Notice that the $X_{\alpha}, X_{\bar\alpha}$ <u>do not belong to the algebra</u> of su(3) but to its complexified algebra. However, <u>the combination</u> (5) does belong to Lie(SU(3)).

(*) This choice corresponds to the choice of <u>a complex structure</u> in G/H [8]. As known from [12,13], others choices are possible : this will be explained in detail elsewhere [14].

II.1. <u>Standard parametrization of the coset space G/H.</u> -

The Coleman, Wess and Zumino parametrization [15] of the coset space G/H (hereafter referred to as C.W.Z.) will then be given, in these complex coordinates, by a unitary matrix $U[\phi,\bar{\phi}]$:

$$U[\phi,\bar{\phi}] = \exp(\phi^{\alpha} X_{\alpha} + \bar{\phi}^{\bar{\alpha}} X_{\bar{\alpha}}) \qquad (6)$$

The left action of an element $g \in G$ on the representative U is

$$g\, U[\phi,\bar{\phi}] = U[\phi',\bar{\phi}']\, h[\phi,\bar{\phi}; g] \qquad (7)$$

where $h \in H$ is the transformation depending on g needed to obtain a $U[\phi',\bar{\phi}']$ of the form (6). The transformed field ϕ' depends on ϕ <u>and $\bar{\phi}$</u> and the isometries will not be holomorphic ones.

This can be traced back to equ. (6) where only a definite combination of X_{α}, $X_{\bar{\alpha}}$ belongs to $\mathfrak{G}$. If in (6) one could forget the relation $\bar{\phi}^{\bar{\alpha}} = (\phi^{\alpha})^{\dagger}$, one would "divide" the representative matrix $U[\phi,\bar{\phi}]$ by $\exp(\bar{\phi}^{\bar{\alpha}} X_{\bar{\alpha}})$ and parametrize the coset space by fields ϕ^{α} only. This remark then leads us to consider the complex extension of the algebras $\mathfrak{G}$ and $\mathfrak{H}$ ($\mathfrak{G}^c$ and $\mathfrak{H}^c$ respectively). As will be shown in the next subsection, this use of complex algebras solves the problem of finding a complex parametrization of G/H in which the isometries are holomorphic.

II. 2 <u>B.K.M.U. parametrization of the coset space SU(3)/((U(1)xU(1))</u>

In this example, one goes from su(3) algebra to sl(3,C) :

$$\mathfrak{G}^c \equiv \begin{pmatrix} \alpha^1 & \eta^3 & \eta^1 \\ \varepsilon^3 & \alpha^2 & \eta^2 \\ \varepsilon^1 & \varepsilon^2 & \alpha^3 \end{pmatrix} \qquad \begin{matrix} \alpha^i, \varepsilon^i, \eta^i \text{ are complex numbers} \\ \text{with } \alpha^1 + \alpha^2 + \alpha^3 = 0 \end{matrix} \qquad (8)$$

We divide the generators of $\mathfrak{G}^c - \mathfrak{H}^c$ into two sets : the upper triangular $\eta^{\alpha} X_{+,\alpha}$ and the lower triangular $\varepsilon^{\alpha} X_{-,\alpha}$. One easily verifies that the $X_{+,\alpha}$ are positively charged generators, with respect to the U(1) charge $Y = H_1 + H_2$:

$$[Y, X_{+,\alpha}] = q_{\alpha} X_{+,\alpha} \qquad q_2 = q_3 = 2,\ q_1 = 4$$

(their hermitian conjugates $X_{-,\alpha}$ being negatively charged).

The set { $\mathfrak{H}^c$, X_+ } of non negatively charged generators span a subalgebra $\tilde{\mathfrak{H}}$ of $\mathfrak{G}^c$. Let $\tilde{H}$ be the corresponding subgroup ($\tilde{H} \supset H^c$). The coset space $G^c/\tilde{H}$ can be described *à la C.W.Z.* by three complex fields ϕ^{α} :

$$\xi(\phi) = \exp(\phi^1 X_{-,1})\exp(\phi^2 X_{-,2})\exp(\phi^3 X_{-,3}) = \begin{pmatrix} 1 & 0 & 0 \\ \phi^3 & 1 & 0 \\ \phi'^1 & \phi^2 & 1 \end{pmatrix} \qquad (9)$$

$$\text{where } \phi'^1 = \phi^1 + \phi^2\phi^3$$

and will shortly be proved to be homeorphic to G/H [13].

The action of an element $g \in G \subset G^c$ on $\xi(\phi)$ is

$$g\,\xi(\phi) = \xi(\phi')\,\tilde{h}[\phi;g] \qquad (10)$$

where the transformation $\tilde{h} \in H$ <u>is now independent of $\bar{\phi}$</u> and <u>is uniquely fixed</u> by the form (9) of $\xi(\phi)$ and $\xi(\phi')$. The infinitesimal version of (10) writes :

$$\xi(\phi+\delta\phi) = \underbrace{\begin{pmatrix} 1 & -\bar{\varepsilon}^3 & -\bar{\varepsilon}^1 \\ \varepsilon^3 & 1 & -\bar{\varepsilon}^2 \\ \varepsilon^1 & \varepsilon^2 & 1 \end{pmatrix}}_{g} \xi(\phi) \underbrace{\begin{pmatrix} \alpha^1 & \eta^3 & \eta^1 \\ 0 & \alpha^2 & \eta^2 \\ 0 & 0 & \alpha^3 \end{pmatrix}}_{\tilde{h}^{-1}[\phi;g]} \qquad (11)$$

One obtains :

$$\alpha^1 = 1 + \bar{\varepsilon}^1\phi'^1 + \bar{\varepsilon}^3\phi^3\,, \quad \alpha^2 = 1 + \bar{\varepsilon}^2\phi^2 - \bar{\varepsilon}^3\phi^3 - \bar{\varepsilon}^1\phi^2\phi^3\,, \quad \alpha^1\alpha^2\alpha^3 = 1$$

$$\eta^1 = \bar{\varepsilon}^1\,, \quad \eta^2 = \bar{\varepsilon}^2 - \bar{\varepsilon}^1\phi^3\,, \quad \eta^3 = \bar{\varepsilon}^3 + \bar{\varepsilon}^1\phi^2 \qquad (12)$$

and :

$$\begin{aligned} \delta\phi^1 &= \varepsilon^1 - \varepsilon^3\phi^2 + \bar{\varepsilon}^1(\phi^1)^2 + \bar{\varepsilon}^2\phi^1\phi^2 \\ \delta\phi^2 &= \varepsilon^2 + \bar{\varepsilon}^3\phi^1 + \bar{\varepsilon}^2(\phi^2)^2 + \bar{\varepsilon}^1\phi^1\phi^2 \\ \delta\phi^3 &= \varepsilon^3 - \bar{\varepsilon}^2\phi'^1 + \bar{\varepsilon}^3(\phi^3)^2 + \bar{\varepsilon}^1\phi'^1\phi^3 \end{aligned} \qquad (13)$$

II.3 <u>$G^c/\tilde{H}$ homeomorphic to G/H</u>

In some neighbourhood of the origin, any element of G^c can be written as :

$$g = \xi(\phi)\exp(a^\alpha(\phi,\bar{\phi})X_{+,\alpha})\exp(c^i(\phi,\bar{\phi})H_i) \qquad (14)$$

Restricting to $g \in G$, we can express a representative of each class of G/H as

$$U[\phi,\bar{\phi}] = \xi(\phi)\exp(a^\alpha(\phi,\bar{\phi})X_{+,\alpha})\exp(c^i(\phi,\bar{\phi})H_i) \qquad (15)$$

where now c^i, i = 1,2 <u>are real</u>. Moreover c^i and a^α are fixed by the unitarity of U. Equation (15) then proves the homeomorphism between G/H and $G^c/\tilde{H}$ [13]. Here one obtains :

$$U = \begin{bmatrix} \frac{1}{\sqrt{w}} & -\frac{\bar{\phi}^3+\phi^2\bar{\phi}'^1}{\sqrt{vw}} & -\frac{\bar{\phi}^1}{\sqrt{v}} \\ \frac{\phi^3}{\sqrt{w}} & \frac{1+\phi^1\bar{\phi}'^1}{\sqrt{vw}} & -\frac{\bar{\phi}^2}{\sqrt{v}} \\ \frac{\phi'^1}{\sqrt{w}} & \frac{\phi^2-\bar{\phi}^3\phi^1}{\sqrt{vw}} & \frac{1}{\sqrt{v}} \end{bmatrix} \quad \begin{aligned} &\text{with: } v = 1+|\phi^1|^2+|\phi^2|^2, \\ &w = 1+|\phi'^1|^2+|\phi^3|^2, \\ &c^1 = -\tfrac{1}{2}\log w + \tfrac{1}{4}\log v\,, \\ &c^2 = -\tfrac{1}{4}\log v\,. \end{aligned} \qquad (16)$$

II.4. <u>Invariant action and vielbeins</u>

The Lie algebra valued Maurer-Cartan one form $T = U^{-1}dU$ can be decomposed as :

$$T = i\omega^i H_i + e^a X_{-,a} + \bar{e}^{\bar{a}} X_{+,\bar{a}} \qquad (17)$$

Due to equation (15), $e^a = e^a{}_\alpha d\phi^\alpha$ (no $d\bar{\phi}^\alpha$) and from equations (16) one obtains

$$e^1 = \frac{d\phi^1 + \phi^3 d\phi^2}{\sqrt{vw}}$$
$$e^2 = \frac{\sqrt{w}\, d\phi^2}{v} - \frac{\bar\phi^3 + \phi^2 \bar\phi^1}{\sqrt{v}} e^1 \qquad (18)$$
$$e^3 = \frac{\sqrt{v}\, d\phi^3}{w} + \frac{\bar\phi^2 - \phi^3 \bar\phi^1}{\sqrt{w}} e^1$$

Under a group transformation $U' = g\, U\, h^{-1}$ (7) and then $T' = hTh^{-1} + hdh^{-1}$. So e^a transforms homogeneously and the $e^a \bar e^a$ ($a = 1,2,3$, no summation) are invariant. As a consequence, the general $SU(3)/(U(1)\times U(1))$ invariant action depends on 3 positive parameters g_a

$$g_{\alpha\bar\beta} = \eta_{a\bar b}\, e^a{}_\alpha\, \bar e^{\bar b}{}_{\bar\beta} \qquad a,\bar b = 1,2,3 \qquad (19)$$

where $\eta_{a\bar b} = g_a \delta_{ab}$ is the (flat) tangent space metric. Here one gets :

$$g_{\alpha\bar\beta}\, \partial_\mu\phi^\alpha \partial_\mu\bar\phi^\beta = (g_1 - (g_2 + g_3))\,|\partial_\mu\phi^1 + \phi^3\, \partial_\mu\phi^2|^2/(vw) \; +$$
$$\partial_\alpha\partial_{\bar\beta}(g_2 \operatorname{Log} v + g_3 \operatorname{Log} w)\, \partial_\mu\phi^\alpha\partial_\mu\bar\phi^\beta \qquad (20)$$

For $g_1 = g_2 + g_3$ the metric is explicitly a regular Kähler one, the Kähler potentials being given by c^1 and c^2 of last subsection : this is general [13] as it will appear in the next section.

III. CLASSICAL ANALYSIS OF ANY HOMOGENEOUS KÄHLER COSET SPACE G/H

As explained in the introduction, G is compact, connected and semi-simple and the subgroup H is the centralizer of a torus. Let iY_j ($j = 1,2..,k \leqslant r$) be the generators of that torus and S_a the other generators of H. By construction $[Y_j, S_a] = 0$. Whatever a Y-charge $\in$ T ($Y = \sum_{i=1,..,k} c^i Y_i$) be, the antihermitian generators of $\mathfrak{G} - \mathfrak{H}$ (in a suitable unitary representation of G) have not a definite charge. But, if one goes to the complex extensions $\mathfrak{G}^c$ and $\mathfrak{H}^c$, one can separate the generators of $\mathfrak{G}^c - \mathfrak{H}^c$ into two parts : the $X_{+,\alpha}$ and the $X_{-,\alpha} = -(X_{+,\alpha})^\dagger$ having positive and negative Y-charge, respectively. Then, with $\tilde{\mathfrak{H}} \equiv \{iY_j, S_a, X_{+,\alpha}\}$ the coset space(*) $G^c/\tilde H$ is parametrized by the complex fields ϕ^α :

$$\xi(\phi) = \exp \phi^\alpha X_{-,\alpha} \qquad (21)$$

The special form of H (the centralizer of a torus) ensures that there is no generator in

(*) $\tilde H$ is the complex subgroup of G^c whose Lie algebra generators are the non-negatively charged $\{\tilde{\mathfrak{H}}\}$.

$\mathcal{G}^0 - \mathcal{H}^0$ with a vanishing Y-charge, which would make the B.K.M.U. trick impossible. Indeed, in such a case the homeomorphism between $G^0/\tilde{H}$ and G/H would be lost. We recall that the choice of Y is equivalent to a choice of a complex structure in G/H, which is not unique [12-14].

As previously explained (equ. 10) the isometries are holomorphic

$$\xi(\phi') = g\,\xi(\phi)\,\tilde{h}^{-1}(\phi;g). \tag{22}$$

Moreover the infinitesimal non-linear(*) g transformations being written

$$g = \exp\,(\varepsilon^\alpha X_{-,\alpha} + \bar{\varepsilon}^\alpha X_{+,\alpha}) \tag{23}$$

one can use the Hausdorf formula and get the following expression for the non- linear infinitesimal isometries :

$$\delta\phi^\alpha = \varepsilon^\alpha + \sum_{p\geq 1} F^\alpha{}_{\beta\gamma_1\ldots\gamma_p}\,\varepsilon^\beta\phi^{\gamma_1}\ldots\phi^{\gamma_p} + \sum_{p\geq 1} G^\alpha{}_{\bar{\beta}\gamma_1\ldots\gamma_p}\,\bar{\varepsilon}^\beta\phi^{\gamma_1}\ldots\phi^{\gamma_p} \tag{24}$$

Both sums are finite, since all components of the field ϕ carry positive Y-charge.

In the symmetric coset space case [9], the $F^\alpha{}_{\beta\gamma_1\ldots\gamma_p}$ vanish and the $G^\alpha{}_{\bar{\beta}\gamma_1\ldots\gamma_p}$ reduce to the bilinear $\frac{1}{2}R^\alpha{}_{\delta\gamma\bar{\beta}}\phi^\delta$ ($R^\alpha{}_{\delta\gamma\bar{\beta}}$ is the Riemann curvature at the origin $\phi = \bar{\phi} = 0$).

The correspondance between C.W.Z. and B.K.M.U. parametrizations (equ. (15)) is still valid and we now explain why the real functions $c^i(\phi,\bar{\phi})$ are the Kähler potentials of the model ($i = 1,..k$) [13]. From (7)(22) and (15) one gets (the Y_i's are hermitian matrices) :

$$\begin{aligned}U(\phi',\bar{\phi}') &= g\,\xi(\phi)\exp(a^\alpha(\phi,\bar{\phi})\,X_{+,\alpha})\exp(b^a(\phi,\bar{\phi})\,S_a)\exp(c^i(\phi,\bar{\phi})\,Y_i)\,h^{-1}(\phi,\bar{\phi};g)\\ &= \xi(\phi')\,\tilde{h}(\phi;g)\exp(a^\alpha(\phi,\bar{\phi})\,X_{+,\alpha})\exp(b^a(\phi,\bar{\phi})\,S_a)\exp(c^i(\phi,\bar{\phi})\,Y_i)\,h^{-1}(\phi,\bar{\phi};g)\\ &\equiv \xi(\phi')\exp(a'^\alpha(\phi,\bar{\phi})\,X_{+,\alpha})\exp(b'^a(\phi,\bar{\phi})\,S_a)\exp(c'^i(\phi,\bar{\phi})\,Y_i)\end{aligned}$$

With $\tilde{h}(\phi;g) = \exp(\tilde{a}^\alpha(\phi;g)\,X_{+,\alpha})\exp(\tilde{b}^a(\phi;g)\,S_a)\exp(d^i(\phi;g)\;Y_i)$
and $h^{-1}(\phi,\bar{\phi};g) = \exp(\delta^a S_a)\exp(i\lambda^i\,Y_i)$ where δ^a and λ^i are real functions, one gets :

$$c'^i(\phi,\bar{\phi}) = c^i(\phi,\bar{\phi}) + \tfrac{1}{2}(\,d^i(\phi;g) + \overline{d^i(\phi;g)}\,) \tag{25}$$

as the c^i and c'^i are real functions.
Equation (25) means that the $c^i(\phi,\bar{\phi})$ are Kähler potentials. The general expression of the Kähler potential $K(\phi,\bar{\phi})$ is then

$$K(\phi,\bar{\phi}) = \sum_{i=1,..k} g_i\,c^i(\phi,\bar{\phi}) \tag{26}$$

(let us recall that k is the dimension of the torus whose centralizer is H). Notice that the general invariant metric depends on m parameters ($m \geq k$) where m is the number of

---

(*) As usual, the subgroup H is supposed to be linearly realized, and only the non-linear isometries are delicate.

irreducible representations of H contained in G/H (3 in the special case of section II). One can show that the metric $g_{\alpha\bar\beta} = \partial_\alpha\partial_{\bar\beta} K(\phi,\bar\phi)$ is a regular one if all the g_i's are different from 0. As a consequence, G/H is a Kähler manifold, even if the more general invariant metric on the manifold is not Kähler : as indicated in refs. [12,13,16], the supersymmetric extension of such bosonic models will enforce relations among the m coupling constants in order that the metric be explicitly Kähler as required [17,8].

IV. ALL ORDER RENORMALISABILITY OF HOMOGENEOUS KÄHLER BOSONIC σ-MODELS

We have obtained - explicitly in the SU(3)/(U(1)xU(1)) case -, the most general invariant metric on the homogeneous Kähler manifold G/H

$$g_{\alpha\bar\beta} = \sum_{i=1}^{m} g_i \, \eta_{a_i\bar b_i} e^{a_i}{}_\alpha e^{\bar b_i}{}_{\bar\beta} \tag{27}$$

where the sum runs over all the irreducible representations R_i of H contained in G/H, and $\eta_{a_i\bar b_i}$ is the matrix of the H-invariant sesquilinear form in the representation R_i. The expression for the non-linear isometries has been found to be

$$\delta\phi^\alpha = \varepsilon^\alpha + \sum_{p\geq 1} F^\alpha{}_{\beta\gamma_1\ldots\gamma_p} \varepsilon^\beta \phi^{\gamma_1}\ldots\phi^{\gamma_p} + \sum_{p\geq 1} G^\alpha{}_{\bar\beta\gamma_1\ldots\gamma_p} \bar\varepsilon^{\bar\beta} \phi^{\gamma_1}\ldots\phi^{\gamma_p} \tag{28}$$

As the transformation law does not close, we introduce an infinite number of sources (refs. [4b,9]) :

$$\Gamma^0 = \Gamma^{class.}(\phi) + \int d^2x \sum_{n=2,..\infty} (\bar L_{I_n} \phi^{i_1}\ldots\phi^{i_n} + L_{I_n} \bar\phi^{i_1}\ldots\bar\phi^{i_n}) \, , \; I_n = \{i_1,\ldots i_n\}$$

The Ward identities for the non-linear symmetry (28) may be written (on the generating functional Γ for 1PI Green functions)

$$W_\varepsilon \Gamma = W_{\bar\varepsilon} \Gamma = 0$$

$$\begin{aligned} W_\varepsilon\Gamma = \varepsilon^\alpha \int d^2x \, \{ & \frac{\delta\Gamma}{\delta\phi^\alpha} + F^\beta{}_{\alpha\gamma}\phi^\gamma \frac{\delta\Gamma}{\delta\phi^\beta} + \sum_{p\geq 2} F^\beta{}_{\alpha\gamma_1\ldots\gamma_p} \frac{\delta\Gamma}{\delta\phi^\beta} \frac{\delta\Gamma}{\delta\bar L_{\gamma_1\ldots\gamma_p}} \\ & - \bar L_{\alpha_1\alpha_2} [\delta^{\alpha_1}{}_\alpha \phi^{\alpha_2} + \sum_{p\geq 1} F^{\alpha_1}{}_{\alpha\gamma_1\ldots\gamma_p} \frac{\delta\Gamma}{\delta\bar L_{\gamma_1\ldots\gamma_p\alpha_2}} + (\alpha_1 \leftrightarrow \alpha_2)] \\ & - \sum_{n\geq 3} \bar L_{\alpha_1\ldots\alpha_n} [\delta^{\alpha_1}{}_\alpha \frac{\delta\Gamma}{\delta\bar L_{\alpha_2\ldots\alpha_n}} + \sum_{p\geq 1} F^{\alpha_1}{}_{\alpha\gamma_1\ldots\gamma_p} \frac{\delta\Gamma}{\delta\bar L_{\gamma_1\ldots\gamma_p\alpha_2\ldots\alpha_n}} \\ & + (\alpha_1 \to \alpha_2 \to \cdots \to \alpha_n)] + \mathcal{O}(\bar\phi, \frac{\delta\Gamma}{\delta L}) \frac{\delta\Gamma}{\delta\bar\phi} \} \end{aligned} \tag{29}$$

Write the general solution as $\Gamma = \Gamma^0 + \Gamma^1$ (it gives the stability of the classical action as well as, due to the use of symmetry preserving dimensional regularization, the structure of one-loop divergences). Due to power counting, Γ^1 has the general structure

$$\Gamma^1 = \int d^2x \left[\Lambda(\partial_\mu\phi, \partial_\mu\bar{\phi}, \phi, \bar{\phi}) + \sum_{n=2,..,\alpha} (\bar{L}_{I_n} T^{I_n}(\phi,\bar{\phi}) + L_{I_n} \bar{T}^{I_n}(\phi,\bar{\phi}))\right] \qquad (30)$$

Extracting terms in $\bar{L}_{I_n}$ from the linearized Ward identity satisfied by Γ^1, we get the following equations for the tensors $T^{I_n}(\phi,\bar{\phi})$

$$\delta_\varepsilon T^{i_1 i_2} = \varepsilon^\alpha \Big[F^{i_1}{}_{\alpha\gamma} T^{i_2\gamma} + \sum_{p\geq 2} F^{i_1}{}_{\alpha\gamma_1\ldots\gamma_p} (T^{\gamma_1\ldots\gamma_p i_2} - \phi^{i_2} T^{\gamma_1\ldots\gamma_p}) + (i_1 \leftrightarrow i_2)\Big] \qquad (31.a)$$

$$\delta_\varepsilon T^{i_1\ldots i_n} = \varepsilon^\alpha \Big[\Big[(\delta^{i_1}{}_\alpha T^{i_2\ldots i_n} + F^{i_1}{}_{\alpha\gamma} T^{\gamma i_2\ldots i_n} + \sum_{p\geq 2} F^{i_1}{}_{\alpha\gamma_1\ldots\gamma_p} (T^{\gamma_1\ldots\gamma_p i_2\ldots i_n} - \phi^{i_2}\ldots\phi^{i_n} T^{\gamma_1\ldots\gamma_p})\Big] + (i_1 \to i_2 \to \cdots \to i_n)\Big] \quad , \quad n\geq 3 \qquad (31.b)$$

In the symmetric case, the right hand side of (31.a) vanishes. One can then use the fact that there is no dimension zero monomial of the fields (no derivatives) invariant under the non linear symmetry to get $T^{i_1 i_2} = 0$. Then, by recurrence, all tensors $T^{i_1\ldots i_n}$ vanish (ref.[9]).

In this homogeneous case, one uses the Y-charge operator. The parametrization has been choosen such that any tensor component $T^{i_1\ldots i_n}$ has a well defined Y-charge. These tensor components can then be ordered according to their charge, regardless to the number of indices they carry. ε^α has a positive charge, and thus one sees from (31.a-b) that the variation of a component with a definite charge contains only components with lower charges. Then, starting from the lowest charged component, one obtains its invariance and then its vanishing. One can then prove by induction on the Y-charge that all tensors $T^{i_1\ldots i_n}$ vanish. <u>The symmetry is not renormalized</u> in that coordinates.

With regard to $\Lambda(\partial_\mu\phi, \partial_\mu\bar{\phi}, \phi, \bar{\phi})$ it should be invariant under the non linear symmetry (28) : the solution is known and allows <u>only coupling constants $\eta_{a\bar{b}}$ renormalization, but no field renormalization</u>.

This proves the multiplicative renormalizability of homogeneous Kähler bosonic non linear σ-models, with no field renormalization in this nice B.K.M.U. parametrization of the coset space G/H. (Of course, a soft mass term has to be added for quantization, but this does not change the result.). The essential ingredients were homogeneity and the existence of a U(1) charge Y that separates ε^α and $\bar{\varepsilon}^\alpha$.

V. CONCLUDING REMARKS

Using the B.K.M.U. parametrization of homogeneous Kähler coset spaces G/H which leads to holomorphic isometries, we have been able to prove the all-order renormalizability of the corresponding bosonic non linear σ-models. Moreover, in that coordinates the fields are unrenormalized. A regulator free treatment *à la B.R.S.* could be done along the lines of ref.[7].

Supersymmetric extensions of these models are possible and this restricts the number of arbitrary coupling constants. Of course, dimensional regularization is no longer usable for renormalizability proofs.

We emphasize that, although homogeneous spaces have, for a given dimension, less stringent geometries than symmetric spaces (for instance, when going from CP^3 = SU(4)/U(3) to SU(3)/(U(1)xU(1)), both describing three complex fields, the isometry group get restricted from SU(4) to SU(3)), it was still possible to renormalize the corresponding σ-models.

References

[1] J. Goldstone, Nuov. Cim. 19 (1961) 165

[2] S. Helgason "Differential geometry, Lie groups, and symmetric spaces" (Academic Press, 1978)

[3] H. Eichenherr and M. Forger,Comm. Math. Phys. 82 (1981) 227 and references therein

[4] a) E. Brézin, J. Zinn Justin and J.-C. Le Guillou, Phys. Rev. D14(1976)2615

b) G. Bonneau and F. Delduc,, Nucl. Phys. B266 (1986) 536

[5] G. Valent, Phys. Rev. D30 (1984) 774

[6] A.V. Bratchikov and I.V. Tyutin, Theor. Math. Phys. 66 (1986) 238

[7] A. Blasi,"B.R.S. renormalization of O(n+1) non-linear σ-model ", these proceedings

[8] G. Bonneau, "Kähler geometry and supersymmetric non-linear σ-models : an introduction", these proceedings

[9] F. Delduc and G. Valent, Nucl. Phys. B253 (1985) 494

[10] a) H.C. Wang, Amer. Jour. Math. 76 (1954) 1,A. Borel and F. Hirzebruch, Amer. Jour. Math. 80 (1958) 458, especially page 501 and chapter IV

b) A. Borel, Proc. Nat. Acad. Sci. USA 40 (1954) 1147

[11] M. Bordemann, M. Forger and H. Römer, Comm. Math. Phys. 102 (1986) 605

[12] M. Bando, T. Kuramoto, T. Maskawa and S. Uehara, Phys. Lett. 138B (1984) 94 and Progr. Theor. Phys. 72 (1984) 313

[13] K. Itoh, T. Kugo and H. Kunitomo,Nucl. Phys. B263 (1986) 295

[14] G. Bonneau, F. Delduc and G. Valent, in preparation

[15] S. Coleman, J. Wess and B. Zumino, Phys. Rev. 177 (1969) 2239

[16] C.L. Ong, Phys. Rev. D31 (1985) 3271

[17] B. Zumino, Phys. Lett. 87B (1979) 203

Nonlinear Field Renormalizations in the Background Field Method

K.S. Stelle*

TH Division
CERN
CH-1211 Geneva 23
Switzerland

ABSTRACT

The use of the background field method in intrinsically nonlinear theories such as σ-models requires nonlinear field renormalizations of the quantum fields that cannot be deduced from divergent graphs with external background lines only. We show how these necessary renormalizations are to be derived, thus allowing computation to arbitrary loop order.

1. Introduction

The background field method is a useful computational tool in quantum field theories that allows one to compute radiative corrections while maintaining manifestly the symmetries of the theory under consideration [1]. In this article, which is based upon work done together with with P.S. Howe and G. Papadopoulos [2],we explain how the method must be applied in the case of intrinsically nonlinear theories such as σ-models. Nonlinear σ-models give rise to new problems in their quantization because maintaining covariance in the perturbative quantum theory requires a nonlinear split between the background and the quantum fields. Although σ-models are renormalizable in two space-time dimensions [3], they are not simply multiplicatively renormalizable. In particular, since the scalar fields of the model have dimension zero, there is nothing to stop them from acquiring arbitrary functional field renormalizations. Because of this, the renormalization of a σ-model in the background field method highlights features of the method that were not brought to the fore in the case of four dimensional gauge theories.

It should be noted that maintaining background symmetries manifest in the quantum formalism is not a substitute for BRS invariance. In Yang-Mills theories, BRS invariance

* On leave of absence from the Blackett Laboratory, Imperial College, London SW7, England.

must be used to establish multiplicative renormalizability, whether one is performing background field quantization [4] or following the classical covariant quantization procedure. BRS techniques also turn out to be necessary in the case of models without internal symmetries, such as σ-models taking their values in target manifolds without isometries. In this case, the background field method has the task of maintaining manifest the invariance under reparameterization of the background scalar fields. BRS methods are needed in this case to control the relation between the background, quantum and total fields. Indeed, similar BRS techniques were used in the original proof of renormalizability of the σ-models in quantization about a constant background [3].

It may be useful to illustrate the issues involved in the context of four-dimensional ϕ^4 theory. The action is

$$S = \int d^4x \Big(\frac{1}{2}(\partial_\mu\phi)^2 - \frac{1}{2}m^2\phi^2 - \frac{\lambda}{4!}\phi^4\Big) \tag{1.1}$$

and the generating functional for connected Green's functions $W[J]$ is defined by

$$e^{iW[J]} = \int \mathcal{D}\phi \exp i(I[\phi] + \textstyle\int J\phi) \tag{1.2}$$

The generating functional for $1PI$ graphs is related to W by a Legendre transformation

$$W[J] = \Gamma[\phi] + \int J\phi \tag{1.3}$$

where the argument of Γ is the vacuum expectation value of the quantum field ϕ in the presence of the source J. We now split the total field ϕ into a background field $\varphi(x)$ and a quantum field $\pi(x)$

$$\phi(x) = \varphi(x) + \pi(x) \tag{1.4}$$

and define a new functional [5] $\bar{W}[\varphi, J]$ by

$$e^{i\bar{W}[\varphi,J]} = \int \mathcal{D}\pi \; exp \; i(S[\varphi + \pi] + \textstyle\int J\pi). \tag{1.5}$$

Evidently

$$\bar{W}[\varphi, J] = W[J] - \int J\varphi. \tag{1.6}$$

Further, we define a new Γ-functional by taking the Legendre transformation with respect to J only,

$$\bar{\Gamma}[\varphi, \pi] = \bar{W}[\varphi, J] - \int J\pi. \tag{1.7}$$

It is not difficult to show using the trivial shift symmetry $\delta\varphi(x) = \eta(x)$, $\delta\pi(x) = -\eta(x)$ that $\bar{\Gamma}$ depends only on $\varphi + \pi$, i.e.

$$\frac{\delta\bar{\Gamma}[\varphi, \pi]}{\delta\varphi(x)} = \frac{\delta\bar{\Gamma}[\varphi, \pi]}{\delta\pi(x)} \tag{1.8}$$

and that

$$\bar{\Gamma}[\varphi, 0] = \Gamma[\varphi]. \tag{1.9}$$

(1.9) is the key equation; it states that the standard $1PI$ functional can be computed by calculating the $1PI$ background functional with no external quantum π lines.

So far, all our considerations have been formal and we have not taken into account the effects of renormalization. Expanding out the interaction term in the action, one finds

$$\frac{\lambda}{4!}\phi^4 = \frac{\lambda}{4!}\left\{\varphi^4 + 4\varphi^3\pi + 6\varphi^2\pi^2 + 4\varphi\pi^3 + \pi^4\right\} \tag{1.10}$$

and in principle these various vertices could be renormalized differently. Now of course this doesn't happen because of the linear-splitting Ward Identity (1.8); furthermore (1.8) also tells us that the wave-function renormalizations for φ and π are actually the same,

$$Z_\varphi = Z_\pi. \tag{1.11}$$

Thus, the Ward identity (1.8) has the consequence that the counterterms are functionals of the total field ϕ, and may be deduced from graphs with no external quantum lines. In particular, the renormalization of the various vertices involving the quantum field is performed by renormalizing λ as deduced from diagrams with only external background lines and then substituting the corresponding bare λ into the right hand side of (1.10). The renormalization of the quantum field π can also be deduced from graphs with only external background lines, as expressed in (1.11), but in fact these multiplicative renormalizations cancel out in graphs with no external quantum lines. This can clearly be seen diagramatically: the factors of Z_π cancel between the propagators and vertices.

In the case of nonlinear theories such as σ-models, the split between background and quantum fields needs to be nonlinear in order to maintain reparameterization covariance. This will give a more complicated Ward identity than (1.8), requiring BRS techniques for its proof. The result will be that the counterterms are not simply functionals of the total field, and additional quantum field renormalizations will be necessary.

2. The Nonlinear σ-Model

Let Ξ be two-dimensional spacetime and M be a Riemannian manifold with metric g. Then the σ-model field is a map $\phi : \Xi \to M$ represented in local coordinates by $\phi^i(x)$. The Lagrangian for the model is

$$L = \frac{1}{2} g_{ij}(\phi)\, \partial_\mu \phi^i \partial^\mu \phi^j. \tag{2.1}$$

The quantization of general σ-models in two space-time dimensions based on the Lagrangian (2.1) has been discussed from a number of different points of view. Friedan [3] considered fluctuations of the σ-model fields about a *constant* background, and then proceeded with the standard approach to quantization. Within this framework, he proved the renormalizability of the theory. An important element in this approach was the proof that the counterterms for a given constant background are geometrical expressions related to those for nearby constant backgrounds in a natural way. The other approaches to the quantization of σ-models have made use of various forms of the background field method [6,7]. This is the most convenient way to carry out renormalization calculations, and counterterms have been derived using the method in a number of theories at low loop orders. The implications of the background field method at higher loop orders in σ-models have not so far been investigated, however. This is the issue to which we now turn.

The action $I = \int d^2x\ L$ is not invariant under any symmetries for a general target manifold M, but it is reparameterization invariant in the sense that L has the same form in any coordinate chart. This can be rephrased slightly: let f be a diffeomorphism of M onto itself; then it induces a new map, $\phi' = f \circ \phi$, and we have

$$I[f_* g,\ \phi'] = I[g, \phi] \tag{2.2}$$

or, infinitesimally,

$$I[g - \mathcal{L}_v g, \phi + v] = I[g, \phi] \tag{2.3}$$

where v is the vector field generating the diffeomorphism and $\mathcal{L}_v$ denotes the Lie derivative. Evidently, a diffeomorphism only induces a symmetry of I, $I[\phi'] = I[\phi]$, if it is an isometry, $\mathcal{L}_v g = 0$. The standard generating functional of connected Green's functions is defined as before

$$e^{iW[J]} = \int \mathcal{D}\phi\ exp\ i \left\{ I[\phi] + \int d^2x\ J_i(x)\ \phi^i(x) \right\} \tag{2.4}$$

but the source term clearly spoils reparameterization invariance.

Formally, the effect of a diffeomorphism is given by

$$e^{iW_{g-\mathcal{L}_v g}[J]} = \int \mathcal{D}\phi\ exp\ i \left\{ I[g, \phi] + \int J \cdot (\phi + v) \right\}. \tag{2.5}$$

Since coupling the source to a function of ϕ leads to the same S-matrix, one concludes that the latter is reparameterization invariant and that models with metrics which are related by a diffeomorphism are physically equivalent.

It is possible to avoid spoiling reparameterization invariance in the Green's functions by the source term if one uses an unconventional source term [3] $\int d^2x\ h(x; \phi(x))$. The functional

$$e^{iW[h]} = \int \mathcal{D}\phi\ exp\ i \left\{ I[g, \phi] + \int h(x; \phi) \right\} \tag{2.6}$$

will be reparameterization invariant provided that $h(x;\phi(x))$ is defined to transform as a scalar. Note that the explicit functional dependence of h on x^μ and $\phi^i(x^\mu)$ means that $\int h(x;\phi)$ is equivalent to a sum of an infinite number of source terms coupling to all powers of the quantum field. Diffeomorphic metrics yield completely equivalent functionals of the form (2.6) since

$$W_{g-\mathcal{L}_v g}[h-\mathcal{L}_v h]=W_g[h]. \tag{2.7}$$

Thus, the nonlinear sigma model is really a theory of an equivalence class of models defined by diffeomorphic metrics. In practice, renormalization calculations will be performed for the conventional functional (2.4) with a single source coupled to the quantum field, but the counterterms can then easily be transformed into forms appropriate to (2.6). For the time being, we will concentrate on renormalizing the functional (2.4), but will return to (2.6) later on.

In proceding to quantize the theory using the background field method, we must now split the total field into background and quantum parts. However, a straightforward linear background-quantum split, $\phi^i=\varphi^i+\pi^i$, does not lead to a manifestly covariant formalism, since π^i cannot be interpreted as a vector. In order to achieve manifest covariance, it is therefore necessary to use a nonlinear split and this can be based on geodesics [7]. Let $\Phi^i(s)$ be an interpolating field with

$$\Phi^i(0)=\varphi^i \ ; \ \left.\frac{d\Phi^i}{ds}\right|_{s=0}=\xi^i \ \text{and} \ \Phi^i(1)=\phi^i \tag{2.8}$$

that satisfies the geodesic equation

$$\frac{d^2\Phi^i}{ds^2}+\Gamma^i_{jk}\frac{d\Phi^j}{ds}\frac{d\Phi^k}{ds}=0. \tag{2.9}$$

Then we can solve (2.9) with the initial conditions of (2.8) to get

$$\phi^i=\varphi^i+\pi^i \ ; \ \pi^i=\xi^i+\chi^i(\varphi,\xi) \tag{2.10}$$

where

$$\chi^i=-\sum_{n=2}^{\infty}\frac{1}{n!}\Gamma^i_{j_1\dots j_n}\xi^{j_1}\dots\xi^{j_n} \tag{2.11}$$

and

$$\Gamma^i_{j_1\dots j_n}=\tilde{\nabla}_{(j_1}\dots\tilde{\nabla}_{j_{n-2}}\Gamma^i_{j_{n-1},j_{n-2})}(\varphi) \tag{2.12}$$

In (2.12) $\tilde{\nabla}$ indicates that the covariant derivative is to be taken with respect to the lower indices only. In this way of splitting, the quantum field is taken to be ξ^i, the tangent vector to the geodesic $\Phi^i(s)$ at $s=0$. Now, ξ has a geometrical interpretation: it is a cross-section of the bundle over Ξ obtained by pulling back the tangent bundle of M with the background field φ, $\xi \in \Gamma[\varphi^*(TM)]$. This fact ensures the covariance of the expansion. To expand the Lagrangian one sets [8]

$$L(s) = \frac{1}{2} g_{ij}(\Phi(s))\, \partial_\mu \Phi^i(s)\, \partial^\mu \Phi^j(s) \tag{2.13}$$

so that

$$\begin{aligned} L(\phi) = L(1) &= \sum_{n=0}^{\infty} \frac{1}{n!} \frac{d^n L(s)}{(ds)^n}\bigg|_{s=0} \\ &= \sum_{n=0}^{\infty} \frac{1}{n!} (\nabla_s)^n L(s)\bigg|_{s=0} \end{aligned} \tag{2.14}$$

where ∇_s is the covariant derivative along the curve $\Phi^i(s)$. The series (2.14) is easily evaluated using the formulae

$$\begin{aligned} &\nabla_s \partial_\mu \Phi^i = \nabla_\mu \frac{d\Phi^i}{ds} \equiv \partial_\mu \frac{d\Phi^i}{ds} + \partial_\mu \Phi^k \Gamma^i_{kj}(\Phi) \frac{d\Phi^j}{ds} \quad , \quad \nabla_s g_{ij} = 0, \\ &\nabla_s \frac{d\Phi^i}{ds} = 0 \quad , \quad [\nabla_s, \nabla_\mu] X^k = \frac{d\Phi^i}{ds} \partial_\mu \Phi^j R^k_{\ell ij} X^\ell, \end{aligned} \tag{2.15}$$

where in the last equation X^i is an arbitrary vector. Hence, all the vertices derived in this expansion involve tensorial functionals of the background metric $g_{ij}(\varphi)$. If we introduce the split (2.10) into $W[J]$ and drop the term $\int J \cdot \varphi$, we obtain

$$\begin{aligned} e^{i\bar{W}[\varphi,J]} &= \int \mathcal{D}\pi \; \exp\, i(I[\phi] + \textstyle\int J_i \pi^i) \\ &= \int \mathcal{D}\xi \; \exp\, i\left(I[\phi] + \textstyle\int J_i(\xi^i + \chi^i)\right). \end{aligned} \tag{2.16}$$

Now (2.16) is not quite what we want since the source is still coupled to the (non-covariant) function χ^i, so we define a new functional

$$e^{iW[\varphi,J]} = \int \mathcal{D}\xi \; exp\; i\left(I[\phi] + \textstyle\int J_i \xi^i\right). \tag{2.17}$$

The Feynman Rules for this functional will be manifestly covariant, but as we shall shortly see, it is not enough to compute $1PI$ graphs with no external lines if one is to determine the counterterms necessary for higher loop calculations from lower loop graphs.

3. The Nonlinear Splitting Ward Identity

The action

$$I[\phi] \equiv I[\varphi, \xi] = I[\varphi + \pi] \tag{3.1}$$

is invariant under the obvious symmetry

$$\begin{aligned} \delta\varphi^i &= \eta^i(x) \\ \delta\pi^i &= -\eta^i(x) \end{aligned} \tag{3.2}$$

and this leads to a simple Ward Identity for the linear splitting $\bar{\Gamma}[\varphi, \pi]$ functional:

$$\frac{\delta\bar{\Gamma}}{\delta\varphi^i} = \frac{\delta\bar{\Gamma}}{\delta\pi^i} \tag{3.3}$$

with the obvious solution

$$\bar{\Gamma}[\varphi, \pi] = \bar{\Gamma}[\varphi + \pi]. \tag{3.4}$$

Now, we can reformulate (3.2) in terms of transformations# of φ and ξ:

$$\begin{aligned} \delta_\eta\varphi^i &= \eta^i \\ \delta_\eta\xi^i &= F^i{}_j(\varphi, \xi)\eta^j \end{aligned} \tag{3.5}$$

where the functional $F^i{}_j$ is determined by the requirement that ϕ^i be invariant, i.e.

$$\delta_\eta\phi^i = \eta^j\delta_j\phi^i + \delta'_j\phi^i F^j{}_k\eta^k = 0 \tag{3.6}$$

where

$$\delta_i \equiv \frac{\delta}{\delta\varphi^i} \quad , \quad \delta'_i \equiv \frac{\delta}{\delta\xi^i} \tag{3.7}$$

(note that since the $\phi \leftrightarrow (\varphi, \xi)$ relation is local in x^μ, we could have used ordinary partial derivatives and ordinary summation here). It is easy to derive the identity

$$\delta_{[j}F^i{}_{k]} + F^\ell{}_{[j}\delta'_\ell F^i{}_{k]} = 0 \tag{3.8}$$

with the aid of which it can be verified that the transformations (3.5) are Abelian. However, they are nevertheless nonlinear and this fact requires that they be handled with care at the quantum level, since the transformations themselves will require renormalization. To study

From now on, we use DeWitt notation, so that an index "i" does double duty as a tensor index and a spacetime point x^μ. Summation over indices includes integration over spacetime.

the associated Ward Identities it is therefore convenient to consider the associated $B.R.S.$ transformations. We introduce a ghost field $c^i(x)$ and define

$$\begin{aligned} s\varphi^i &= c^i \\ s\xi^i &= F^i{}_j c^j \\ sc^i &= 0. \end{aligned} \tag{3.9}$$

Then $s^2 = 0$ by virtue of (3.8). To obtain the Ward Identity it is necessary to modify the action by including $s\xi^i$ coupled to an anticommuting source L_i. However, since $F^i{}_j c^j$ has power counting weight zero it will mix with all other possible dimension zero operators. To allow for this operator mixing we therefore define

$$\Sigma = I + L_{\alpha i} N^i_\alpha \tag{3.10}$$

where

$$\begin{aligned} N^i_\alpha &= s\Lambda^i_\alpha \qquad \alpha = 0, 1, \ldots \infty \\ \Lambda^i_0 &\equiv s\xi^i \quad ; \quad L_{0i} \equiv L_i. \end{aligned} \tag{3.11}$$

The set $\{\Lambda^i_\alpha\}$ are all possible dimension zero vectorial functions of φ and ξ, this set being sufficient as we shall see. If we assign dimension zero and ghost number $n_g = 1$ to c then the L's have dimension 2 and ghost number -1. It is clear that

$$s\Sigma = 0 \ , \tag{3.12}$$

since $s\phi = 0$ and $s^2 = 0$. This can be rewritten as

$$s\Sigma = c^i \frac{\delta\Sigma}{\delta\varphi^i} + \frac{\delta\Sigma}{\delta L_i} \frac{\delta\Sigma}{\delta\xi^i} = 0 \tag{3.13}$$

since

$$\delta\xi^i = \frac{\delta\Sigma}{\delta L_i} \ .$$

At the quantum level, the structure of the $1PI$ functional $\Gamma[\varphi, \xi, L, c]$ is determined by the background reparameterization invariance and by the shift Ward identity

$$c^i \frac{\delta\Gamma}{\delta\varphi^i} + \frac{\delta\Gamma}{\delta L_i} \frac{\delta\Gamma}{\delta\xi^i} = 0 \ . \tag{3.14}$$

Reparameterization invariance is realized linearly on the quantum fields, yeilding the reparameterization Ward identity

$$\mathcal{L}_v g_{ij} \frac{\delta\Gamma}{\delta g_{ij}} - v^i \frac{\delta\Gamma}{\delta\varphi^i} - \partial_j v^i \xi^j \frac{\delta\Gamma}{\delta\xi^i} = 0 \ . \tag{3.15}$$

Assuming the existence of an ultraviolet regularization scheme that preserves reparameterization invariance#, the linearity of (3.15) in Γ implies that the same relation will be satisfied by the divergent part of Γ, Γ^D. In the present case of a purely bosonic σ-model, dimensional regularization can be used as an invariant ultraviolet regulator. Thus, the divergences will be manifestly invariant under reparameterizations of the target manifold.

The shift Ward identity (3.14) is nonlinear in Γ and consequently requires some care in its analysis. We wish to subtract the divergences so that the renormalized Γ, Γ^r, continues to satisfy (3.14). As in the Yang-Mills case, we proceed order by order in $\hbar$. For example, the one-loop divergences satisfy

$$D_\Sigma \Gamma^D_{(1)} \equiv c^i \frac{\delta \Gamma^D_{(1)}}{\delta \varphi^i} + \frac{\delta \Gamma^D_{(1)}}{\delta L_i} \frac{\delta \Sigma}{\delta \xi^i} + \frac{\delta \Sigma}{\delta L_i} \frac{\delta \Gamma^D_{(1)}}{\delta \xi^i} = 0 \tag{3.16}$$

The solution to (3.16) is given by

$$\Gamma^D_{(1)} = G[\phi] + D_\Sigma X[\varphi, \xi, L, c] \; , \tag{3.17}$$

where $G[\phi]$ is a reparameterization invariant functional of the total field ϕ, and X is an arbitrary functional with ghost number -1 and dimension zero (remembering that it is an integrated functional). That (3.17) solves (3.16) follows from the fact that $sG[\phi] = 0$ and by virtue of the fact that the operator D_Σ is nilpotent,

$$D_\Sigma D_\Sigma = 0 \tag{3.18}$$

as may be verified directly. That it is the most general solution can be proved using similar arguments to the Yang-Mills case [9]. X has the general form

$$X = Z_{\alpha\beta} L_{\alpha i} \Lambda^i_\beta \tag{3.19}$$

for some (infinite) constants $Z_{\alpha\beta}$. Expanding out (3.17) yields

$$\Gamma^D_{(1)} = G[\phi] - Z_{\alpha\beta} L_{\alpha i} N^i_\beta + Z_{0\alpha} \Lambda^i_\alpha \frac{\delta \Sigma}{\delta \xi^i} \; . \tag{3.20}$$

It follows from the reparameterization Ward identity (3.15) that $\Gamma^D_{(1)}[\varphi, \xi, c = L = 0]$ is covariant, so that the Λ^i_α are indeed vectors as claimed, and in addition $G[\phi]$ must be of the form $-\frac{1}{2} T_{ij} \partial_\mu \phi^i \partial_\mu \phi^j$. To summarize, the one loop divergences comprise metric divergences

Two dimensional massless theories also require infrared regularization. This can be done in a reparameterization and shift (3.5) invariant way by including a potential $m^2 V(\phi)$, where V is a scalar function of the total field. The function $V(\phi)$ will also have to be renormalized, e.g. at the one-loop order by a term proportional to $D^i D_i V$. Since these renormalizations are proportional to m^2, they are clearly distinguishable from the other ultraviolet divergences with which we are chiefly concerned.

(from G), nonlinear ξ renormalizations $\left(Z_{0\alpha}\Lambda^i_\alpha \frac{\delta\Sigma}{\delta\xi^i}\right)$ and multiplicative renormalizations of the infinite set of sources $\{L_{\alpha i}\}$. The sources $\{L_{\alpha i}\}$ are needed only in the proof of renormalizability and will ultimately be set to zero. Thus, the important renormalizations are those of the metric and of the quantum field ξ.

One can iterate the above procedure loop by loop. In the end, the renormalized action has the form

$$\Sigma^{(r)}[\varphi,\xi,L,c] = I^0[\varphi,\xi^0] + L^0_{\alpha i} N^i_\alpha(\varphi,\xi^0) \tag{3.21}$$

where I^0 includes the metric counterterms, i.e. $g_{ij} \to g^0_{ij} = g_{ij} + \Sigma T_{ij}$ and

$$\begin{aligned} L^0_{\alpha i} &= L_{\beta i} Z_{\beta\alpha} \\ Z_{0\alpha}\Lambda^i_\alpha(\varphi,\xi^0) &= \xi^i \ . \end{aligned} \tag{3.22}$$

The proof that (3.21) satisfies the splitting Ward identity

$$c^i \frac{\delta\Sigma^{(r)}}{\delta\varphi^i} + \frac{\delta\Sigma^{(r)}}{\delta L_i}\frac{\delta\Sigma^{(r)}}{\delta\xi^i} = 0 \tag{3.23}$$

follows from the observation that

$$s^0\Sigma^{(r)} = 0 \tag{3.24}$$

where

$$\begin{aligned} s^0\varphi^i &= c^i \\ s^0\xi^{0i} &= F^i{}_j(\varphi\xi^0)c^j = \frac{\delta\Sigma^{(r)}}{\delta L^0_i} \\ s^0 c^i &= 0 \end{aligned} \tag{3.25}$$

because

$$N^i_\alpha(\varphi,\xi) = s\Lambda^i_\alpha(\varphi,\xi) \Rightarrow N^i_\alpha(\varphi,\xi^0) = s^0\Lambda^i_\alpha(\varphi,\xi^0) \ . \tag{3.26}$$

Using (3.25), (3.24) is

$$c^i \frac{\delta\Sigma^{(r)}}{\delta\varphi^i} + \frac{\delta\Sigma^{(r)}}{\delta L^0_i}\frac{\delta\Sigma^{(r)}}{\delta\xi^{0i}} = 0 \ . \tag{3.27}$$

If one then changes variables from (φ,ξ^0) to (φ,ξ) and uses (3.22), one readily sees that (3.27) implies (3.25). Strictly speaking, we should show that the solution (3.21) is unique; this can be done along the same lines as the Yang-Mills case as discussed, for example, in Ref. [9].

To summarize then, the consequences of the nonlinear splitting Ward Identity are given in equations (3.21) and (3.22); in addition to the metric renormalizations, there are nonlinear renormalizations of the quantum field ξ^i that are not derivable from expanding out the metric counterterms. Furthermore, they do not cancel out in higher loop graphs and therefore must be taken into account. In the next section we give an algorithm for computing them.

4. Computational Algorithm

In this section we show how to renormalize the functional

$$e^{iW[\varphi,J]} = \int \mathcal{D}\xi e^{i(I[\varphi,\xi]+\int J\xi)} \,. \tag{4.1}$$

From the preceding section we know that the renormalized functional is given by

$$e^{iW[\varphi,J]} = \int \mathcal{D}\xi e^{i(I^0[\varphi,\xi^0]+\int J\xi)} \,, \tag{4.2}$$

which follows from (3.21) and (3.22) upon setting the $L_{\alpha i}$ to zero. The task is to compute the metric contributions and the (nonlinear) quantum wave function renormalizations. This could, of course, be done straightforwardly by computing all $1PI$ graphs with both external quantum and background lines. To do this would be to violate the spirit of the background field method, however, in which the effective action $\Gamma[\varphi]$ is computed from graphs with no external quantum lines. Nonetheless, as we have shown, the nonlinear renormalizations of the quantum field must be taken into account in order to correctly subtract the theory. Moreover, these renormalizations cannot be deduced from the divergent parts of $\Gamma[\varphi]$ alone.

In order to deduce the necessary renormalizations of g_{ij} and ξ^i without computing graphs with external ξ lines, we consider instead of (4.1) the generalized functional (2.6),

$$e^{iW[h]} = \int \mathcal{D}\xi e^{i\{I[\varphi,\xi]+\int h(x,\phi)\}} \tag{4.3}$$

The functional (4.3) can be renormalized in the same way as (4.1) since the modified source is a functional of the total field, but the wave-function renormalizations cancel out since $W[h]$ is essentially the vacuum functional for the modified action $I + \int h$. So (4.3) is renormalized by

$$\begin{aligned} g_{ij} \to g^0_{ij} &= \mu^\epsilon\Big(g_{ij} + T_{ij}(g)\Big) \\ h \to h^0 &= \mu^\epsilon\Big(h + H(g,h)\Big) \end{aligned} \tag{4.4}$$

where the source counterterms H are linear in h because h has dimension two. As with the metric counterterms, the h counterterms can be classified according to their conformal weights under the scalings $g_{ij} \to \lambda^{-1} g_{ij}$, $h \to \lambda^{-1} h$. For example, the one-loop counterterms have conformal weight zero, so there are only two possibilities, $\nabla^i\nabla_i h$ and Rh. The latter does not occur since the Feynman rules involve only derivatives of h.

In terms of the nonlinear $\varphi - \xi$ split, h has the expansion

$$h(x,\ \phi(x)) = h(x,\ \varphi(x)) + \sum_{n=1}^{\infty} \frac{1}{n!}\,\xi^{i_1}\ldots\xi^{i_n}\,(\nabla_{i_1}\ldots\nabla_{i_n}h)_{\phi=\varphi} \tag{4.5}$$

If we choose the condition

$$\nabla_{(i_1}\ldots\nabla_{i_n)}h\,\Big|_{\phi=\varphi} = 0 \qquad n \neq 1 \tag{4.6}$$

then the source term $\int h$ reduces to a conventional source term as in (4.1). When this is done after renormalization, it is necessary to perform a quantum field redefinition in order to recast the renormalized integrand into the form of (4.2).

To see how this works, consider the bare source h^0 which occurs in the renormalized functional

$$e^{iW^{(r)}[h]} = \int \mathcal{D}\xi \exp\, i\Big(I^0[\varphi,\ \xi] + \textstyle\int h^0(x;\phi)\Big)\ ; \tag{4.7}$$

one finds

$$\mu^{-\epsilon}h^0 = h + \frac{1}{4\pi\epsilon}\nabla^i\nabla_i h + \ell \geq 2\ loop\ terms\ . \tag{4.8}$$

Expanding now h^0 using (4.5) and (4.6), one finds

$$\mu^{-\epsilon}h^0 = h_i\Big(\xi^i + \sum_{\ell=1}^{\infty} X^i_{(\ell)}(\varphi,\ \xi)\Big)\ ,\quad h_i = \nabla_i h\,\Big|_{\phi=\varphi}\ , \tag{4.9}$$

where the $X^i_{(\ell)}$ are covariant expressions corresponding to ℓ loops involving all powers in the quantum field ξ. These arise from the renormalization of h upon the imposition of (4.6). Note that $X^i_{(\ell)}$ is determined entirely by the ℓ-loop contribution to h^0 in (4.8); $X^i_{(\ell)}$ contains terms of arbitrary order in the quantum field ξ and is a power series in the regulator ϵ^{-1} up to order $\epsilon^{-\ell}$. For example, from the one-loop contribution to h^0 given in (4.8) one obtains

$$X^i_{(1)} = -\frac{1}{4\pi\epsilon}\Big(\frac{2}{3}R^i_j\,\xi^j + \frac{1}{2}\nabla_j R^i_k\xi^j\xi^k - \frac{1}{12}\nabla^i R_{jk}\xi^j\xi^k + O(\xi^3)\Big)\ . \tag{4.10}$$

We can regain the form (4.2) by changing variables in the functional integral so that the source h_i couples to ξ^i, with the resulting bare quantum field ξ^{0i} in the action given by

$$\xi^{0i} = \xi^i - X^i_{(1)}(\varphi,\xi) - X^i_{(2)}(\varphi,\xi) + X^j_{(1)}\frac{\delta}{\delta\xi^j}\,X^i_{(1)} + \ell \geq 2\ loop\ terms\ . \tag{4.11}$$

Again, we emphasize that the expression for the bare quantum field ξ^{0i} at a given loop order ℓ is derived from the renormalization of h at loop orders up to ℓ given in (4.8).

In order to compute the renormalization of I and h in (4.7), it is convenient in practice to consider a more general functional

$$e^{iW[\varphi,h_i,J]} = \int \mathcal{D}\xi \exp i\left(I[\varphi,\xi] + \textstyle\int h + \int J_i\xi^i\right)\ . \tag{4.12}$$

This functional obviously reduces on the one hand to (4.3) for $J_i = 0$ and on the other hand to (4.1) for $h = 0$. In order to renormalize (4.12), we require metric, h and ξ renormalizations. Since h has dimension two, its presence does not affect the renormalization of ξ^i, which has dimension zero. Thus, by our previous discussion, the renormalization of ξ^i is given in terms of the renormalization of h by (4.11). We also know from the results of section 3 that the renormalization of h in (4.12) is given by functionals of the total field $\phi(\varphi, \xi)$, so it may be deduced from diagrams with no external ξ lines. Hence, by calculating with the general functional (4.12), we may deduce the renormalization of the quantum field ξ^i via the renormalization of h from diagrams with no external quantum lines, and then set $h = 0$ to obtain the renormalized (4.2).

5. Conclusion

In this article, we have analyzed the general structure of the counterterms for a nonlinear σ-model defined on an arbitrary Riemannian manifold. In addition to the expected counterterms which are functionals of the total field ϕ^i, there are additional nonlinear field renormalizations of the quantum field ξ^i which must be performed even if one wishes to calculate only diagrams without external quantum lines. These nonlinear renormalizations of ξ^i can be calculated from the renormalization of a generalized source $h(x; \phi^i(x))$, which is treated as if it were a potential for the σ-model.

The renormalizations of the quantum field that we have discussed above are required for the correct subtraction of subdivergences in all higher loop orders. Since these renormalizations are nonlinear, they do not simply cancel out as do the multiplicative quantum field renormalizations in four-dimensional Yang-Mills theories [4,5]. At the two-loop level, the subdivergences that are removed by these renormalizations are proportional to the classical equations of motion for the background. At higher loop orders, one will encounter second and higher order variations of the action with respect to the background fields. Properly taking account of these subdivergences is necessary for the separation of ultraviolet from infrared divergences. For example, if one uses dimensional regularization plus a potential incorporating a mass m as an infrared regulator, at the two loop order there are simultaneous ultraviolet and infrared subdivergences proportional to $\frac{1}{\epsilon}\ell n\left(\frac{m^2}{\mu^2}\right)$ that are cancelled by the quantum field renormalization [2].

For simplicity, we have been concerned in this paper only with σ-models without fermionic fields and without torsion. All of the above considerations are of equal importance in more general cases. Recent work on the renormalization of nonlinear σ-models with torsion [10] has confirmed the general structure described here.

References

[1] B.S. de Witt, in *Quantum Gravity 2*, eds. C. J. Isham, R. Penrose and D.W. Sciama (Clarendon Press), 449; G. 't Hooft in *Proc. 12th Winter School in Theoretical Physics in Karpacz, Acta Univ. Wratisl.* no. 38, (1975); D.G. Boulware, *Phys. Rev.* **D23** (1981) 389.

[2] P.S. Howe, G. Papadopoulos and K.S. Stelle, Institute for Advanced Study preprint, Dec. 1986.

[3] D. Friedan, *Phys. Rev. Lett.* **45** (1980) 1057; *Ann Phys.* **163** (1985) 318.

[4] H. Kluberg-Stern and J.B. Zuber, *Phys. Rev.* **D12** (1975) 482, 3159.

[5] L. Abbot, *Nucl. Phys.* **B185** (1981) 189.

[6] J. Honerkamp, *Nucl. Phys.* **B36** (1972) 130.

[7] L. Alvarez-Gaumé, D.Z. Freedman and S. Mukhi, *Ann. Phys.* **134** (1981) 85.

[8] S. Mukhi, *Nucl. Phys.* **B264** (1986) 640.

[9] S. Joglekar and B.W. Lee, *Ann. Phys. (N.Y.)* **97** (1976) 160.

[10] C.M. Hull and P.K. Townsend, University of Cambridge D.A.M.T.P. preprint, Mar. 1987.

KÄHLER GEOMETRY AND SUPERSYMMETRIC NON-LINEAR σ- MODELS : AN INTRODUCTION

Guy BONNEAU

Laboratoire de Physique Théorique et Hautes Energies,

Université Paris VII, Tour 24, 2 place Jussieu 75251 PARIS CEDEX 05, FRANCE

Abstract. The necessary and sufficent conditions for a supersymmetric extension of a bosonic non-linear σ-model to exist are reviewed. The framework for the perturbative analysis of such models is sketched with emphasis on some delicate points. These are exemplified on the "proof" of all-orders finiteness of hyper-Kähler supersymmetric non-linear σ-models.

I. INTRODUCTION

The importance of Kähler geometry for supersymmetric theories was stressed in 1979 by B. Zumino [1]. He studied the SUSY extensions of a bosonic non-linear σ-model whose fields take values in a complex Kähler manifold and showed that if the supersymmetry is N=1 for four-dimensional space-time, it turns to N=2 in two space-time dimensions. Necessary and sufficient conditions for extended supersymmetry in two space-time dimensions where later on given by L.Alvarez Gaumé and D.Z.Freedman [2] and we shall review them in the first part of this talk. Perturbation theory for supersymmetric non-linear σ-models will then be sketched (background field method with normal coordinates, D.Friedan [3] renormalizability in the space of metrics, one loop calculations and all-orders finiteness conjectures) with emphasis on some delicate - and, to our mind, controversial - points.

Let us also mention another application of Kähler geometry : in general relativity, it is well known that the geometry of self-dual gravitational instantons is that of an hyper-Kähler manifold (see G. Valent contribution to this worshop [4]).

Our interest for Kähler geometry comes from our attempts at defining a physical field theory through characterizations other than the usual Ward identities linked to isometries (symmetric spaces or homogeneous manifolds [5]).Such characterizations could be S matrix properties such as non production [6] , scale invariance or geometric properties such as Kähler or hyper-Kähler structure.

II. SUPERSYMMETRIC EXTENSIONS OF BOSONIC NON-LINEAR σ-MODELS

This section closely follows the discussion made in ref [2]. Starting with the bosonic action

$$I[\phi] = \frac{1}{2}\int d^2x\, g_{ij}[\phi]\, \partial_\mu \phi^i\, \partial_\mu \phi^j \qquad (1)$$

where the n fields ϕ^i's are understood as coordinates on an n-dimensional real Riemannian manifold $\mathcal{M}$ whose metric is $g_{ij}[\phi]$, an N=1 supersymmetric extension will be :

$$I[\phi,\psi] = \frac{1}{2}\int d^2x\, \{g_{ij}[\phi]\partial_\mu\phi^i\partial^\mu\phi^j + i g_{ij}[\phi]\bar{\psi}^i\gamma^\mu D_\mu\psi^j + \tfrac{1}{6}R_{ijkl}(\bar{\psi}^i\psi^k)(\bar{\psi}^j\psi^l)\} \qquad (2)$$

D_μ is the covariant derivative : $D_\mu\psi^i = \partial_\mu\phi^j\nabla_j\psi^i = \partial_\mu\psi^i + \Gamma^i_{jk}\,\partial_\mu\phi^j\psi^k$. Γ^i_{jk} and R_{ijkl} the Christoffel connection and Riemann curvature corresponding to the metric g_{ij}. The supersymmetry transformations

$$\delta^{(1)}\phi^i = \bar{\varepsilon}\,\psi^i, \qquad \delta^{(1)}\psi^i = -i\,\partial\!\!\!/\,\phi^i\,\varepsilon - \Gamma^i_{jk}(\bar{\varepsilon}\psi^j)\,\psi^k \qquad (3)$$

commute with coordinate reparametrizations of $\mathcal{M}$. In ref [2], it is shown that there is a second supersymmetry leaving the action (2) invariant, and satisfying SUSY algebra (*) without central charges

$$\{Q^{(a)}, \bar{Q}^{(b)}\} = 2\,\delta^{ab}\, P\!\!\!/ \qquad (4)$$

<u>If and only if</u> a tensor $f^i{}_j[\phi]$ exists with the properties :

$$f^i{}_j\, f^j{}_k = -\,\delta^i{}_k \qquad (5.a)$$

$$f^i{}_j\, g_{ik}\, f^k{}_l = g_{jl} \qquad (5.b)$$

$$\nabla_k f^i{}_j = 0 \qquad (5.c)$$

This second supersymmetry is completely fixed in function of $f^i{}_j[\phi]$:

$$\delta^{(2)}\phi^i = \bar{\varepsilon}\,(f^i{}_j\psi^j), \qquad \delta^{(2)}(f^i{}_j\psi^j) = -i\,\partial\!\!\!/\,\phi^i\,\varepsilon - \Gamma^i_{jk}(\bar{\varepsilon}\, f^j{}_m\psi^m) f^k{}_n\psi^n$$

and i has the same expression as (3) with ψ^i changed to $f^i{}_j\psi^j$.

We now interpret equations (5) from a geometric point of view [7] :

- Equation (5.a) $\Leftrightarrow$ <u>$f^i{}_j$ is an almost complex structure</u>. Complex coordinates Z^α, $\bar{Z}^{\bar{\alpha}}$ can be defined locally that diagonalize the almost complex structure : $f^\alpha{}_\alpha = i$, $f^{\bar{\alpha}}{}_{\bar{\alpha}} = -i$ ($\alpha,\bar{\alpha} = 1,\ldots,m$, where $n=2m$).

- Equations (5.a+b) $\Leftrightarrow$ <u>g_{ij} is hermitian with respect to $f^i{}_j$</u> $\Leftrightarrow$ the Riemannian manifold <u>$(\mathcal{M}, g_{ij})$ is an almost hermitian manifold</u>. In the complex coordinates $Z^\alpha, \bar{Z}^{\bar{\alpha}}$, $g_{\alpha\beta} = g_{\bar{\alpha}\bar{\beta}} = 0$.

(*) If the manifold is irreducible, it is shown in [2] that any new fermionic invariance of the action is necessarily a supersymmetry satisfying the algebra (4).

An antisymmetric tensor is then defined : $f_{ij} = g_{ik} f^k{}_j = -f_{ji}$ and a 2-form, called the Kähler form :

$$\Omega = \tfrac{1}{2} f_{ij}\, d\phi^i \wedge d\phi^j = i g_{\alpha\bar\beta}\, dZ^\alpha \wedge d\bar Z^{\bar\beta} \tag{6}$$

All these are local properties whose global extension is possible if, and only if, $f^i{}_j$ satisfies the integrability condition that the Nijenhuis tensor vanishes

$$N_{ij}{}^k = f^l{}_i (\partial_l f^k{}_j - \partial_j f^k{}_l) - (i \leftrightarrow j) = 0. \tag{7}$$

- Equations (5.a+7) $\Leftrightarrow$ <u>$\mathcal{M}$ is a complex manifold.</u> Then, in each open set one can choose complex coordinates such that, in the intersection of 2 charts, the coordinate systems Z^α, Z'^α are holomorphically related :

$$Z'^\alpha = f^\alpha(Z) \qquad , \bar Z'^{\bar\beta} = \bar f^{\bar\beta}(\bar Z)$$

- Equations (5.a+b +7) $\Leftrightarrow$ <u>$(\mathcal{M}, g_{ij})$ is an Hermitian manifold.</u> The property $g_{\alpha\beta} = g_{\bar\alpha\bar\beta} = 0$ is preserved by an analytic change of coordinates.

- Equations (5.a+b+c) $\Leftrightarrow$ <u>$\mathcal{M}$ is a Kähler manifold</u> $\Leftrightarrow$ $d\Omega = 0$. This is a global notion ((5.c)$\Rightarrow$ $N_{ij}{}^k = 0$) which is a strong restriction on $\mathcal{M}$: it means that, in a coordinate frame adapted to the hermitian structure $f^i{}_j$, there exists a <u>Kähler potential</u> $K(Z,\bar Z)$ such that

$$g_{\alpha\bar\beta} = \partial^2 K(Z,\bar Z)/(\partial Z^\alpha \partial \bar Z^{\bar\beta}). \tag{8}$$

$K(Z,\bar Z)$ is defined up to a Kähler transformation :

$$K(Z,\bar Z) \rightarrow K(Z,\bar Z) + f(Z) + \bar f(\bar Z) \tag{9}$$

If two covariantly constant complex structures exist, satisfying the Clifford algebra that results from supersymmetry (equ.4) :

$$f^{(a)} f^{(b)} + f^{(b)} f^{(a)} = -2\delta^{ab} \qquad a,b = 1,2$$

then $f^{(3)} = f^{(1)} f^{(2)}$ is also a covariantly constant complex stucture and we get N=4 supersymmetry. We then have three covariantly constant complex structures $f^{(a)}$, a=1,2,3, satisfying the SU(2) (quaternionic) relations :

$$f^{(a)i}{}_k f^{(b)k}{}_j = -\delta^{ab} \delta^i{}_j + \varepsilon^{abc} f^{(c)i}{}_j \tag{10}$$

and $\mathcal{M}$ is called <u>an hyper-Kähler manifold.</u>

The following table summarizes the known results on supersymmetric extensions of bosonic non-linear σ-models defined on a Riemannian manifold $(\mathcal{M}, g_{ij})$. (A 4-dimensional N-supersymmetric theory gives through dimensional reduction to d=2 a 2N-supersymmetric one : this justifies the d = 4 column of the table). We emphasize that this table shows an equivalence between a geometric property of a manifold and some physical symmetry of a (supersymmetric non-linear σ -) model built on that manifold. For general bosonic models, such equivalence does not seem to exist, and, in view of quantization one might hope to take such geometrical property as the definition of the theory.

extended SUSY \ space time dimension	d = 2	d = 4
N = 1	no restriction on $\mathcal{M}$	$\mathcal{M}$ is Kähler
N = 2	$\mathcal{M}$ is Kähler	$\mathcal{M}$ is hyper-Kähler
N = 4	$\mathcal{M}$ is hyper-Kähler	No extension exists (bosonic sector needs spin 1)

With this in mind, we now discuss the perturbative approach to these supersymmetric non-linear σ- models.

III. PERTURBATION THEORY FOR SUPERSYMMETRIC NON-LINEAR σ-MODELS

The background field method is inherent in the perturbative study of generalized non-linear σ- models *à la Friedan* [3].

We also add a mass term to the action by hand, since here we are not concerned with the infra-red behaviour.

III.1 Background field method with normal coordinates

The usual background field splitting $\phi^i = f^i_{back} + Q^i$ leads to a non covariant formalism as Q^i is not a vector field under reparametrizations. So, in 1972, J.Honerkamp introduced normal coordinates [8]

$$\phi^i = f^i_{back} + \zeta^i + \chi^i(f_{back}, \zeta) \qquad (11)$$

where the quantum field ζ^i is the tangent vector to the geodesic $\Phi^i(t)$ at t=0 :

$$\Phi^i(0) = f^i_{back}\ ,\ \Phi^i(1) = \phi^i\ ,\ d\Phi^i(t)/dt\,\big|_{t=0} = \zeta^i \qquad (12)$$

The resulting Feynman rules will be manisfestly covariant and, due to power counting, the divergences of the theory could be compensated for by covariant metric counterterms $T^{(1)}_{ij}$ [3] :

$$g_{ij}[\phi] \rightarrow g^{bare}_{ij}[\phi] = g_{ij}[\phi] + (\hbar/\varepsilon)\, T^{(1)}_{ij}[g] + \ldots\ldots \qquad (13)$$

(dimensional regularization, compatible with reparametrization invariance is used here). Finiteness proofs for supersymmetric non-linear σ-models will be based upon an analysis of the $T^{(1)}_{ij}$ allowed in perturbation theory.

A few comments are in order :

C.1. the polynomial character of the divergences, used to write equ. (13), supposes substraction of subdivergences : these ones, being necessarily ambiguous, have to be precisely fixed through normalisations conditions. Here, in the absence of isometries for the general Riemannian metric g_{ij}, an infinite number of such normalisation conditions - i.e of *a-priori* physical parameters - is necessary. This non renormalisability is usually circumvented by invoking minimal schemes : we emphasize that this relies heavily on a definite regularization and hides the difficulty. A correct approach would be to prove, as was done in other examples with an infinite number of parameters [9], that only a finite number of them are physical ones.

C.2. As shown by B. De Wit and M.T.Grisaru [10], when non linearly realized symmetries are present, the "on-shell counterterms" are not necessarily symmetric, even if a symmetry preserving regulator existed. As a consequence, when studying extended supersymmetry, one should devise a background field splitting such that the transformations of the quantum field under reparametrizations and extended supersymmetry are linear : this is not so simple [11].

C.3. Moreover, in d = 2 non-linear σ-models where the canonical dimension of the field vanishes, non-linear field renormalizations are involved[9]. The usual argument -that quantum field renormalizations are unnecessary in the background field method -seems difficult to maintain, and indeed, it has been proved in a recent calculation (see K.S. Stelle contribution to this worshop [12]) that background field counterterms are not sufficient to compensate for non local higher loop divergences ($(1/\varepsilon)\log m^2/\mu^2$ terms in a two-loop calculation).

C.4. Quantization of such supersymmetric theories relies heavily on the existence of a supersymmetry preserving regulator. The common practice is to use dimensional reduction [13a)] : all the spinor -or supersymmetric covariant derivatives - algebra is done in d = 2 dimensions and, after, momenta are analytically continued in the complex d-plane. Unfortunately, this method suffers from mathematical inconsistencies [13b),c)] and one cannot rely upon it. In superfield calculations, convergence improvements occur only after such manipulations are down on divergent integrals, and thus they suffer from the very ambiguities that make normalisation conditions necessary.

III. 2. One-loop calculations and all-orders finiteness conjectures in the supersymmetric case.

One-loop on-shell divergences are unambiguous and proportionnal to the Ricci tensor $R_{ij}[g]$ (refs.[9,3]). Two- and three-loop calculations, with minimal substraction and dimensional reduction, indicate (but for comments C.1 and C.4) that no new divergent contribution appears after the one-loop order. This led to the conjectures - and claimed proofs - that N = 1 supersymmetric non-linear σ-models have only a one loop divergence or, less ambitiously, that N= 2 Ricci flat supersymmetric σ- models are all-orders finite [14]. All these "proofs" suffer from the difficulties previously mentioned, and a recent four-loop calculation[15], done under the same hypothesis, shows that they are uncorrect.

As N = 2 Ricci flat supersymmetric non-linear σ-models are discussed in C.N. Pope contribution to this workshop [16], we now present and comment the less ambitious conjecture of all-order finiteness for supersymmetric non-linear σ-models built on hyper-Kähler manifolds[17](hyper-Kähler implies Ricci flatness [2]).

The equivalence discussed in section II between N = 2 (resp. N = 4) supersymmetry and the Kähler (resp. hyper-Kähler) character of the metric, plus the hypothesis of a supersymmetry preserving regulator, imply that the counterterms should be Kähler (resp. hyper-Kähler).

The Kähler character of the counterterms $T_{\alpha\bar\beta}$ means that they are of the form $T_{\alpha\bar\beta} = \partial_\alpha \partial_{\bar\beta} S[Z,\bar Z]$ where, *à-priori*, $S[Z,\bar Z]$ is not a globally defined function. However, in refs.[17a),11], arguments are given which indicate that, except at the one loop order where $R_{\alpha\bar\beta} = \partial_\alpha \partial_{\bar\beta} \mathrm{Log} \det |g|$, $S[Z,\bar Z]$ is a globally defined function. N = 2 supersymmetry, plus Ricci-flatness of the metric to get rid of the one loop contribution, then insures that the Kähler form Ω (equ. (6)) stays in the same cohomology class in $H^{1,1}(\mathcal{M})$.

On the other hand, the hyper-Kähler character of the counterterms means that the Ricci flatness of the metric is preserved in higher orders :

$$N = 4 \text{ SUSY} \quad \Rightarrow \quad R_{\alpha\bar\beta}[g] = R_{\alpha\bar\beta}[g + T] = 0 \qquad (14)$$

Then the Ricci form $J = iR_{\alpha\bar\beta}\, dZ^\alpha \wedge d\bar Z^{\beta}$ is the same for metrics g and g+T whose Kähler forms Ω and Ω' are in the same cohomology class. The manifold being supposed compact (**) and connected, the uniqueness theorem of Calabi[18] asserts that these Kähler forms are the same and, as a consequence, $T_{\alpha\bar\beta}$ vanishes, to all-orders of perturbation theory.

Q.E.D.

We hope to have clearly pointed out the delicate points of such a proof (see comments C.1 and C.4). Another "proof" of finiteness of N=4 supersymmetric σ-models exists, based upon quantization in harmonic superspace [19]. It of course suffers from the same difficulties.

(**)Here, as a consequence of Ricci flatness, the first Chern class vanishes. Concerning compactness, we do not expect that perturbative results sould depend on this mathematical necessary hypothesis (see also [17b]).

IV. CONCLUDING REMARK

If we expressed doubts on some "proofs" in the litterature on supersymmetric non-linear σ-models, however we think that something should be true in these up to five-loop order calculations [15]. As in the early days of Q.E.D. recalled by D. Maison in his introductory talk [20], when there was no satisfactory way to get rid of infinities (renormalization theory being not yet at hand), new methods are probably necessary to explain these "experimental" results.

REFERENCES

[1] B. Zumino, Phys.Lett.87B(1979) 203

[2] L. Alvarez-Gaumé and D.Z. Freedman, Com.Math.Phys.80 (1981) 443

[3] D.H.Friedan, Phys.Rev.Lett. 45 (1980) 1057 and Ph.D. thesis, August 1980, published in Ann.Phys. 163 (1985) 318

[4] G. Valent, Methods in hyper-Kähler σ-models building,these proceedings

[5] G. Bonneau, Renormalization of bosonic non-linear σ-models built on compact homogeneous manifold, these proceedings

[6] G. Bonneau and F. Delduc, Nucl.Phys.B250 (1985) 561

[7] a) A detailed introduction to Kähler geometry for physicists can be found in L.Alvarez-Gaumé and D.Z.Freedman lecture at Erice 1980, "Unification of the fundamental particule interactions", eds. S. Ferrara et all., Plenum New York 1980, page 41 ,
b) for a more mathematical point of view, see S.Gallot contribution in "première classe de Chern et courbure de Ricci : preuve de laconjecture de Calabi" Société mathématique de France, Astérisque n°58 (1978)

[8] J.Honerkamp, Nucl.Phys.B36 (1972) 130

[9] a) O. Piguet and K. Sibold, these proceedings
b) G.Bonneau and F. Delduc, Nucl.Phys. B266 (1986) 536

[10] B. de Wit and M.T. Grisaru, Phys. Rev. D20 (1979) 2082

[11] P.S. Howe, G. Papadopoulos and K.S. Stelle, Phys. Lett. 174B (1986) 405

[12] K.S. Stelle, these proceedings

[13] a) W. Siegel, Phys.Lett.84B (1979)193
b) W. Siegel, Phys.Lett.94B (1980) 37
c) L.V. Adveev et all., Phys.Lett. 105B (1981) 272

[14] Unpublished preprints by L. Alvarez-Gaumé and P. Ginsparg, by C.M. Hull ; L.Alvarez-Gaumé, S. Coleman and P. Ginsparg, Comm.Math.Phys. 103 (1986) 423

[15] M.T. Grisaru, A.E.M. van de Ven and D. Zanon, Phys.Lett. 173 B (1986) 423 , M.T. Grisaru, D.I.Kazakov and D. Zanon, HUTP preprint 1987

[16] C.M. Pope, these proceedings

[17] a) L. Alvarez-Gaumé and P. Ginsparg, Comm. Math. Phys.102(1985) 311
b) C.M. Hull, Nucl.Phys. B260 (1985) 182

[18] For a pedagogical review on Calabi-Yau theorems, see the book refered under[7b)]

[19] A. Galperin et all., Class. Quant. Gravity 2 (1985) 617

[20] D. Maison, these proceedings.

METHODS IN HYPERKÄHLER σ MODELS BUILDING

D. OLIVIER, G. VALENT
LPTHE, Université Paris VII, T. 24, 5°étage
2, place Jussieu 75251 PARIS CEDEX 05
FRANCE

1° INTRODUCTION

The study of supersymmetric extensions of bosonic non linear σ models has received increasing attention these last years. The main reason being that they are deeply related, already at the classical level, with the theory of G structures which is by itself a field of interest. Indeed for a four dimensional base manifold the metric which defines the σ model should be Kähler (K) (resp. Hyper-Kähler (HK)) to accomodate for N = 1 (resp N=2) supersymmetries [1] .

At the quantum level, the increase in the number of complex structures linked to extended supersymmetries is commonly believed to give milder ultraviolet divergences [2] . This indicates that if something can "stabilize" at all the G structures this should be supersymmetry.

However the increase of complexity in constructing explicit metrics is significant : for K. ones it is completely solved through the existence of a potential (in holomorphic coordinates), for the H.K. ones no such a general characterization has yet been found.

In view of these remarks the use of N=2 susy in HK building seems therefore an attractive approach, which already led to interesting general results [3] .

Recently an unconstrained N=2, D=4 superfield formalism was constructed [4] : the so called Harmonic Superspace (HSS). In this framework, starting from a given superfield lagrangian one is in principle able to extract out a bosonic sector which must be HK. This raises the hope of a systematic approach to the construction of HK metrics.

Indeed the superfield lagrangians corresponding to Taub-NUT [5], Eguchi-Hanson and the higher dimensional metrics of Calabi [6] are known.

It is the aim of this talk to review the results obtained for Taub-NUT in [5] and to present some generalizations of them [7,8,14].

2° HK METRICS AND GRAVITATIONAL INSTANTONS

Let us first recall some basic results in four dimensional HK metrics, using the notations of [9].

The euclidean distance writes in terms of the vierbein:

$$ds^2 = \sum_{A=0}^{3} e_A^2$$

and the connexion ω_{AB} and curvature R_{AB} result from :

$$de_A + \omega_{AB} \wedge e_B = 0 \qquad \omega_{AB} = -\omega_{BA}$$

$$R_{AB} = d\omega_{AB} + \omega_{AC} \wedge \omega_{CB} \qquad R_{AB} = -R_{BA}$$

The curvature (as well as the connexion) can be splitted using self-duality :

$$R_{AB} = R^+_{AB} + R^-_{AB} \qquad R^\pm_{AB} = \pm \frac{1}{2} \varepsilon_{ABCD} R^\pm_{CD}$$

This splitting can be interpreted in terms of the holonomy algebra whose generators are precisely the R_{AB}. For a general metric the holonomy algebra is so(4) and its generators can be splitted into two subsets R^+_{AB} and R^-_{AB} which generate respectively $su(2)_+$ and $su(2)_-$.

It follows that a metric with self dual curvature ($R^-_{AB} = 0$) has holonomy $su(2)_+ \sim sp(1)$ and is therefore HK, as first obtained in [1]. However such an argument does not tell anything on the complex structures. Here we will present a proof of this result which gives in addition their explicit form.

For any 4 dimensional metric we begin by defining a triplet of 2-forms :

$$\Omega_i(e) = e_0 \wedge e_i - \frac{1}{2}\varepsilon_{ijk}\, e_j \wedge e_k = \frac{1}{2}(F_i)_{\mu\nu}\, dx^\mu \wedge dx^\nu$$

$$i, j, k = 1, 2, 3.$$

It is readily checked that the quaternionic multiplication law holds :

$$(F_i)_\mu{}^\nu\,(F_j)_\nu{}^\lambda = -\delta_{ij}\,\delta_\mu{}^\lambda + \varepsilon_{ijk}\,(F_k)_\mu{}^\lambda$$

and, therefore any 4 dimensional metric exhibits a triplet of (almost) complex structures and it is (almost) hermitian with respect to them.

It can be further shown [7] that they satisfy :

$$d\Omega_i = -2\,\bar{\omega}_{ij} \wedge \Omega_j$$

$$\nabla_\gamma (F_i)_{\mu\nu} = -2\,\bar{\omega}_{ij\gamma}\,(F_j)_{\mu\nu} \qquad \bar{\omega}_{ij} = \bar{\omega}_{ij\mu}\, dx^\mu$$

Now if the metric has self dual curvature, its connexion is a pure gauge. There exists a matrix $M \in SU(2)_-$ such that :

$$\bar{\omega}_{AB} = (M^{-1} dM)_{AB}$$

Then if we rotate the vierbein :

$$e_A \longrightarrow \hat{e}_A = M_{AB}\, e_B$$

we get $\hat{\bar{\omega}}_{AB} = 0$.

It follows that the triplet of complex structures $(\hat{F}_i)_{\mu\nu}$ extracted from $\Omega_i(\hat{e})$ are then covariantly constant establishing the HK character of the metric.

It is interesting to notice that such an HK structure implies Ricci flatness and ensures that any euclidean HK metric is a gravitational instanton.

Let us now turn to the HSS derivation of Taub-NUT.

3° Taub-NUT INSTANTON AND HSS

In this formalism the basic object is the unconstrained superfield q^+ [4] which lives in the analytic N=2 superspace $\{x_A^m, \theta_\alpha^+, \bar\theta_{\dot\alpha}^+, u_i^\pm\}$ and is defined by :

$$q^+ = F^+ + i\theta^+\sigma^m\bar\theta^+ A_m^- + \theta^+\theta^+ M^- + \bar\theta^+\bar\theta^+ N^- + \theta^+\theta^+\bar\theta^+\bar\theta^+ P^{(-3)} + \text{fermions}$$

The price to pay for unconstrained superfield is the existence of infinitely many auxiliary fields which are displayed in the harmonic expansions of F^+, A^-, etc... For instance F^+ writes :

$$F^+ = f^i u_i^+ + \sum_{n=1}^{\infty} a^{i i_1 \dots i_n j_1 \dots j_n}(f,\bar f)\, u^+_{(i i_1 \dots i_n} u^-_{j_1 \dots j_n)}$$

In this expansion the spinor coordinates $f^i (i=1,2)$ and their charge conjugate partners $\bar f^i$ are the physical bosonic fields. They correspond to a super isospin 1/2. The infinite tail of higher isospins must be expressed in terms of the coordinates using the field equations.

For Taub-NUT the action is [5] :

$$\int dx_A\, du\, d^2\theta^+ d^2\bar\theta^+ \left\{ \overline{q^+} D^{++} q^+ + \frac{\lambda}{2}(q^+\overline{q^+})^2 \right\} \qquad (1)$$

(see [4,5] for the notations -the involution $*$ of these references is simplified here to —).

The first attractive feature of this formalism is that the isometries of the bosonic sector can be read off from the superfield lagrangian. Here we have the supersymmetric SU(2) for which f^i and $\bar f^i$ are doublets and the so-called Pauli-Gursey U(1) :

$$U(1)_{PG} \longrightarrow \quad \delta q^+ = i\eta\, q^+ \qquad \delta\overline{q^+} = -i\eta\,\overline{q^+}$$

The second nice feature is that, given the lagrangian, the remaining steps to obtain the metric are all "deductive". Let us describe the HSS algorithm :

1) solve for the field equations :

$$D^{++} q^+ + \lambda q^+ (q^+ \overline{q^+}) = 0$$

and get F^+, A^- in terms of the coordinates f^i, $\bar{f}^i$.

2) integrate over θ^+, $\overline{\theta^+}$ in (1) and get rid of the auxiliary fields M,N,P using their equations of motion. One remains with :

$$\frac{1}{2} \int d^4x \, du \left(\overline{A_m^-} \partial_m F^+ - A_m^- \partial_m \overline{F^+} \right)$$

3) extract out the singlet part from this expression. This gives you an HK metric !

In order to express the final result it is convenient to define :

1) <u>bispinor coordinates</u> : $f^{ia} = \begin{cases} f^i & a=1 \\ \bar{f}^i & a=2 \end{cases}$

where $f^{i,1}$, $f^{i,2}$ are spinors under $SU(2)_S$ and $f^{1,a}$, $f^{2,a}$ are spinors under $SU(2)_{PG}$. These coordinates are constrained by pseudo-reality

$$\overline{f^{ia}} = \varepsilon_{ij} \varepsilon_{ab} f^{jb}$$

2) <u>the vierbein E^{ia}</u> :

with the same pseudo-reality constraint as the coordinates. For Taub-NUT :

$$E^{i1} = \frac{1}{\sqrt{1+\lambda s}} \left\{ \left(1+\frac{\lambda s}{2}\right) df^i - \frac{\lambda}{2} f^i (f_k d\bar{f}^k) - \frac{\lambda}{2} \bar{f}^i (f_k df^k) \right\} \quad (2)$$

$$E^{i2} = \varepsilon^{ij} \overline{E^{j1}} \qquad s = f^i \bar{f}_i$$

The relevant information on Taub-NUT can then be summarized in a triple :

$$\left.\begin{array}{lll} \text{. isometries} & \longrightarrow & SU(2)_S \times U(1)_{P.G.} \\ \text{. distance} & \longrightarrow & ds^2 = E^{ia} E^{jb} \varepsilon_{ij} \varepsilon_{ab} \\ \text{. triplet of closed 2 forms :} & \longrightarrow & \Omega^{ij} = E^{ia} \wedge E^{jb} \varepsilon_{ab} \end{array}\right\} \quad (3)$$

Let us observe that the structure displayed by equations (3) was first obtained by Sierra and Townsend [10] for any N=2 susy σ model using <u>constrained superfields</u>. However the detailed form of the vierbein (2) was found in [5] using HSS.

These relations have such a nice structure, that one may address the following question: to what extent do these relations define Taub-NUT and this, independently of the use of HSS.

Let us try to construct the most general metric satisfying (3) with isometries $SU(2)_S \times U(1)_{PG}$. Its vierbein should write

$$E^{iu} = A(s)\, df^i + B(s)\, \bar{f}^i (f_k\, df^k) + C(s)\, f^i (f_k\, d\bar{f}^k)$$

because of the isometries. The closedness of the triplet Ω^{ij} imposes :

$$\frac{d}{ds}(XY) = 2XYT \qquad X = (A - sB)^2$$

$$\frac{d}{ds} X = -sXYT - X(Y - T) \qquad Y = \frac{B+C}{A-sB} \qquad (4)$$

$$T = \frac{B-C}{A-sB}$$

The function T(s), which remains free, reflects the arbitrariness in the definition of s.

The system (4) with the choice T = 0 leads to the vierbein (2) (provided that we look for a metric which is flat at the origin).

This proves that, at least at the classical level, Taub-NUT is "uniquely" defined by requiring (3) and the isometries $SU(2)_S \times U(1)_{PG}$. "Uniquely" means here up to a reparametrization of s :

$$s \to \quad s' = s f(s) \qquad f(0) = 1$$

4° SYMMETRY BREAKING AROUND Taub-NUT

We have generalized in [8] and [14] respectively both of the previous analyses.

Let us present our results.

1) the HSS analysis was applied to the lagrangian :

$$\mathcal{L} = \overline{q^+} D^{++} q^+ + \frac{\lambda}{2}(q^+ \overline{q^+})^2 + \frac{\xi^{--}}{3}(q^+ \overline{q^+})^3$$

where $\qquad \xi^{--} = \xi^{ij} u_i^- u_j^- \quad , \quad \overline{\xi^{--}} = -\xi^{--}$

induces a symmetry breaking of $SU(2)_S$ down to $U(1)_S$; therefore we have $U(1)_S \times U(1)_{PG}$ as isometries.

It leads to the vierbein

$$E^{i1} = \frac{1}{\sqrt{1+s\tilde{\lambda}}}\left\{\left(1+s\frac{\tilde{\lambda}}{2}\right)df^i - \frac{1}{2}\left(\tilde{\lambda}-\frac{t}{3}\right)\bar{f}^i(f_k df^k) - \frac{s}{6}\bar{\zeta}^i(f_k df^k) - \frac{1}{2}\left(\tilde{\lambda}-\frac{t}{3}\right)f^i(f_k d\bar{f}^k) - \frac{s}{6}f^i(\zeta_k d\bar{f}^k)\right\}$$

where :

$$\zeta^i = \xi^{ij} f_j \quad , \quad t = \xi^{ij} f_i \bar{f}_j \quad , \quad \tilde{\lambda} = \lambda - \frac{2t}{3}$$

and to a triplet of closed 2-forms Ω^{ij} as in (3).

At that stage one has to face the problem of the identification of this HK metric.

In fact GIBBONS [11] pointed out that this metric should be related to the multicenter ones [12]. Indeed using polar coordinates :

$$f^1 = \sqrt{s}\cos\frac{\theta}{2}\, e^{\frac{i}{2}(\psi+\varphi)} \qquad\qquad f^2 = \sqrt{s}\sin\frac{\theta}{2}\, e^{\frac{i}{2}(\psi-\varphi)}$$

and rotating the vector ξ^{--} to the form :

$$\xi^{--} = 2\xi u_1^- u_2^-$$

and defining :

$$V(r) = 1 + \frac{2m}{r} + \frac{2}{3}(2m)^3\xi r\cos\theta \qquad \lambda = \frac{1}{4m^2}$$

$$\vec{\omega} : \quad \vec{\nabla} V = \vec{\nabla}\times\vec{\omega} \qquad r = \frac{s}{2m} \qquad \tau = 2m(\psi+\varphi)$$

the distance derived from E^{ia} becomes :

$$4ds^2 = V^{-1}(d\tau + \vec{\omega}\cdot d\vec{x})^2 + V d\vec{x}\cdot d\vec{x} \qquad r = |\vec{x}|$$

which is the standard form of the multicenter metrics. The lagrangian we started from corresponds to a potential V with one center (corresponding to Taub-NUT) plus a symmetry breaking term with dipolar structure.

2) the "relations (3)" analysis was applied to obtain the explicit form of a metric which generalizes Taub-NUT by lowering its isometries to $SU(2)_s$. The detailed computations are too hairy to be presented here [14] but lead ultimately to the known gravitational instanton of Atiyah and Hitchin [13].

It is interesting to note that this metric has not yet been recovered in the HSS approach in spite of the fact that the generic form of its lagrangian is well known [5] . It appears that the breaking of $U(1)_{PG}$ introduces in the field equations non linearities which are hard to deal with.

In our opinion this second approach may give some help to the HSS analysis.

REFERENCES

[1] ALVAREZ-GAUME L., FREEDMAN D.Z., Commun. Math. Phys. 80, 443 (1981)

[2] For a critical discussion, see the contributions of C. BECCHI and G. BONNEAU to this workshop.

[3] HITCHIN N.J., KARLHEDE A., LINDSTROM U., ROCEK M., Commun. Math. Phys. 108, 535 (1987).

[4] GALPERIN A., IVANOV E., KALITZIN S., OGIEVETSKY V., SOKATCHEV E., Class. Quantum Grav. 1, 469 (1984).

[5] GALPERIN A., IVANOV E., OGIEVETSKY V., SOKATCHEV E., Commun. Math. Phys. 103, 515 (1986).

[6] GALPERIN A. IVANOV E., OGIEVETSKY V., TOWNSEND P.K., Class. Quantum Grav. 3, 625 (1986).

[7] OLIVIER D., VALENT G., preprint PAR-LPTHE 86/22, unpublished.

[8] OLIVIER D., VALENT G., preprint PAR-LPTHE 86/49 to appear in Physics Letters B.

[9] EGUCHI T., GILKEY P., HANSON A., Phys. Rep. 66, 213 (1980).

[10] SIERRA C., TOWNSEND P., Phys. Lett. 124B, 497 (1983).

[11] GIBBONS G.W., Private Communication.

[12] HAWKING S.W., Phys. Lett. 60A, 81 (1977).
GIBBONS G.W., HAWKING S.W., Phys. Lett. 78B, 430 (1978).

[13] ATIYAH M.F., HITCHIN N.J., Phys. Lett. 107A, 21 (1985).
GIBBONS G.W., MANTON N.S., Nucl. Phys. B274, 183 (1986).

[14] OLIVIER D., VALENT G., in preparation.

SIGMA MODEL β-FUNCTIONS AT ALL LOOP ORDERS

C.N. Pope
CERN, CH-1211 Geneva 23
Switzerland

1. Introduction

One of the most remarkable claims, indeed perhaps the remarkable claim, of string theory is that it constitutes a consistent quantum theory, possibly finite, that incorporates general relativity as a low-energy effective limit. This low-energy theory corresponds to Einstein's theory of gravity described by the Lagrangian $\sqrt{-g}R$, together with higher-order terms involving more derivatives, such as $\sqrt{-g}\,(\text{Riem})^2$, etc. One way to investigate these higher-order terms in the effective Lagrangian is to calculate string scattering amplitudes, in which the external lines are chosen to be on-shell gravitons. Such amplitudes have been calculated, for the type IIB and heterotic string theories, with up to four external gravitons[1)], and give rise to terms in the effective action up to quartic in Riemann tensors.

A different approach to determining the low-energy effective action is from the σ-model point of view. The two-dimensional worldsheet action for a closed string coupled to a non-trivial curved background metric g_{ij} is classically invariant under conformal rescalings of the worldsheet metric. Consistency requires that this conformal invariance be preserved at the quantum level, in other words that the worldsheet stress tensor should not develop a trace-anomaly. This is equivalent to the condition that the β-function β_{ij} that describes the renormalization of the target -space metric g_{ij} should vanish. The resulting equation,

$$0 = \beta_{ij} = R_{ij} + \cdots \tag{1}$$

is believed to be equivalent to that obtained from the low-energy effective action for the string.

In this paper we review some results concerning the form of the contributions to β-functions for σ-models in non-trivial backgrounds. Section 2 is concerned with supersymmetric σ-models with curved target-space metrics g_{ij}. The forms of the metric counterterms that can arise are tightly constrained by supersymmetry considerations. A sequence of candidate counterterms at all loop-orders can be constructed[2)]; the four-loop term had already been found by direct calculation[3)], and shown to be equivalent to $(\text{Riem})^4$ term in the type IIB string effective action[2),4),5)]. It is not known whether the higher-loop terms in the sequence actually occur with non-zero coefficients in general, nor whether other independent counterterms can arise.

In section 3, we consider a rather simpler σ-model corresponding to an open bosonic string in flat 26-dimensional space, in the background of a non-zero electromagnetic field $F_{\mu\nu}$. The string couples to the potential A_μ only via a boundary term, and because the problem is therefore essentially one-dimensional, it turns out that the β-function β_μ describing the renormalization of A_μ can be calculated exactly. Although perhaps of limited physical relevance, this provides an interesting toy example in which one could, in principle, study the non-perturbative structure of the quantum theory.

2. Supersymmetric σ-models

The action for the supersymmetric σ-model is

$$I = \int d^2\sigma \Big\{ g_{ij}\, \partial_+ X^i \partial_- X^j + i \lambda_L^a \nabla_+ \lambda_L^a + i \lambda_R^a \nabla_- \lambda_R^a + \tfrac{1}{2} R_{abcd} \lambda_L^a \lambda_L^b \lambda_R^c \lambda_R^d \Big\} , \tag{2}$$

where $g_{ij}(X)$ is the metric on the target manifold M, $\partial_\pm = \partial/\partial\tau \pm \partial/\partial\sigma$, X^i are the co-ordinates on the target space, and λ_L^a, λ_R^a are left and right moving fermions. It is invariant under the supersymmetry transformations

$$\begin{aligned} \delta X^i &= \varepsilon_L \lambda_L^i + \varepsilon_R \lambda_R^i , \\ \delta \lambda_L^i &= -\Gamma^i_{jk} \lambda_L^j \delta X^k + i \varepsilon_L \partial_- X^i , \\ \delta \lambda_R^i &= -\Gamma^i_{jk} \lambda_R^j \delta X^k + i \varepsilon_R \partial_+ X^i , \end{aligned} \tag{3}$$

where $\lambda^a = \lambda^i e_i^a$ and $e_i^a(X)$ satisfies $e_i^a e_j^a = g_{ij}$. This N = 1 supersymmetry is in fact comprised of two independent supersymmetries, of type (1,0) and (0,1), corresponding to the independent parameters ε_L and ε_R, and is often referred to as (1,1) supersymmetry.

If the target manifold M is Kähler, then there is a second, independent, supersymmetry with parameters $(\varepsilon_L', \varepsilon_R')$, in which $(\lambda_L^i, \lambda_R^i)$ in Eq. (3) are replaced by $(J^i{}_j \lambda^j{}_L, J^i{}_j \lambda_R^j)$, where $J^i{}_j$ is the complex structure on M, with $J_{ij} = -J_{ji} = g_{ik} J^k{}_j$

the Kähler form. This N = 2 supersymmetry is further enlarged to N = 4 if M is hyperkähler, the analogous additional supersymmetries occurring for each of the three complex structures $J^{(1)i}{}_j$, $J^{(2)i}{}_j$ and $J^{(3)i}{}_j$.

The action (2) may be written in terms of N = 1 superfields as

$$I = \int d^2\sigma \, d\theta_L \, d\theta_R \, g_{ij}(\phi) \, D_R\phi^i \, D_L\phi^j \, , \tag{4}$$

where $\phi^i = X^i + \theta_L\lambda_L^i + \theta_R\lambda_R^i + \theta_L\theta_R F^i$ and $D_L = \partial_{\theta_L} + i\theta_L\partial_-$, $D_R = \partial_{\theta_R} + i\theta_R\partial_+$. The fermionic co-ordinates θ_L and θ_R are left and right-handed Majorana spinors. In the N = 2 case, where M is a Kähler manifold with Kähler metric g_{ij}, the action may be written in terms of N = 2 superfields as

$$I = \int d^2\sigma \, d^4\theta \; K(\phi) \tag{5}$$

where $K(\theta)$ is the Kähler potential.

The N = 4 (hyperkähler) models are finite[6]. The reason for this is essentially that the order-by-order renormalization of g_{ij} must preserve the hyperkähler condition. In particular, this implies that the metric is Ricci-flat and Kähler, and hence (for a given complex structure) unique. Thus no counterterms can arise.

For the N = 2 (Kähler) models, one can integrate out two of the θ's in Eq. (5), to obtain Eq. (4) with $g_{a\bar{b}} = \partial_a\partial_{\bar{b}}K$ in complex co-ordinate notation. Likewise, counterterms ΔI will take the form $\int d^2\sigma d^4\theta S$, and can be integrated out to give $\Delta g_{a\bar{b}} = \partial_a\partial_{\bar{b}}S$. General arguments based on considering the conformal weights of the counterterms show that, beyond one loop, the functions S must be globally-defined scalars constructed from powers of the Riemann tensor on M and its covariant derivatives[7].

Since all hyperkähler spaces are Kähler, and all Kähler spaces are Riemannian spaces, it follows that the N = 4, N = 2 and N = 1 supersymmetric σ-models can all be viewed as N = 1 models. It therefore follows from the above discussion that all metric counterterms Δg_{ij} must be Riemannian expressions which vanish if g_{ij} is hyperkähler and must take the form $\Delta g_{a\bar{b}} = \partial_a\partial_{\bar{b}}S$ if g_{ij} is Kähler. The problem of constructing candidate counterterms can thus be reduced to finding symmetric tensors T_{ij} built from R_{ijkl}'s and their covariant derivatives that satisfy these requirements.

In order to construct such tensors, it is convenient to employ real rather than complex co-ordinates in the case of Kähler spaces. To do this, we define projectors $\pi^{\pm}{}_i{}^j$:

$$\pi^{\pm}{}_i{}^j = \frac{1}{2}\left(\delta_i{}^j \pm i J_i{}^j\right). \qquad (6)$$

These project onto holomorphic and antiholomorphic indices, so that $V_a \leftrightarrow \pi^+{}_i{}^j V_j$, $V_{\bar a} \leftrightarrow \pi^-{}_i{}^j V_j$. Thus counterterms must take the form $T_{ij} \sim \pi^+{}_{(i}{}^k \pi^-{}_{j)}{}^l \nabla_k \nabla_l S$ in Kähler spaces, i.e.

$$T_{ij} = \nabla_i \nabla_j S + \nabla_{\hat\imath} \nabla_{\hat\jmath} S , \qquad (7)$$

where we use the hat notation[8], defined by $V_{\hat\imath} = J_i{}^j V_j$ for all vector V_i. A candidate counterterm is acceptable if S vanishes when g_{ij} is hyperkähler, and if Eq. (7) can be written without complex structures, i.e. without hats. This last requirement is highly non-trivial.

The first example of an acceptable counterterm that does not vanish in Ricci-flat spaces was found in Ref. 3) for N = 2 σ-models, and it was shown that it in fact occurs with non-zero coefficient at the four-loop order. The scalar S is given by

$$S = R_{ij}{}^{kl} R_{kl}{}^{mn} R_{mn}{}^{ij} - 2 R_{ikjl} R^{kmln} R_m{}^i{}_n{}^j . \qquad (8)$$

It is straightforward to check that S vanishes if g_{ij} is hyperkähler. This follows immediately from the form of the Riemann tensor on hyperkähler spaces, $R_{ijkl} \leftrightarrow \varepsilon_{\alpha\beta}\varepsilon_{\gamma\delta}\Omega_{ABCD}$, where α, β are Sp(1) indices, A, B are Sp(n) indices (dim M = 4n), and $\Omega_{ABCD} = \Omega_{(ABCD)}$. To prove that T_{ij} in Eq. (7) can be written without complex structures, one must show that $\nabla_{\hat\imath} \nabla_{\hat\jmath} S$ can be written without hats. It is a tedious and not entirely straightforward matter to do this from Eq. (8) by repeated use of cyclic and Bianchi identities[8]. One also needs to use the fact that, from the properties of J_{ij}, it follows that $V_{\hat\imath}U^i = -V_i U^{\hat\imath}$, $V_{\hat{\hat\imath}} = -V_i$, and $R_{ijk\hat l} = -R_{ij\hat k l}$. This last property follows from the covariant-constancy of J_{ij}, which also implies that hats can pass freely through covariant derivatives.

There is a much easier way to derive the result that $\nabla_{\hat\imath}\nabla_{\hat\jmath}S$ can be written without complex structures, which also lends itself immediately to a generalization to candidate counterterms at arbitrary loop order. By using a cyclic identity, one can easily see that Eq. (8) may be written as

$$S = 2 R^{\hat{r}_1}{}_{r_2}{}^{\hat{\kappa}_1}{}_{\kappa_2} R^{r_2}{}_{r_3}{}^{\kappa_2}{}_{\kappa_3} R^{r_3}{}_{r_1}{}^{\kappa_3}{}_{\kappa_1} . \tag{9}$$

This admits a natural generalization to an expression of n'th power in Riemann tensors[2],

$$S^{(n)} = 2^{n-2} R^{\hat{r}_1}{}_{r_2}{}^{\hat{\kappa}_1}{}_{\kappa_2} R^{r_2}{}_{r_3}{}^{\kappa_2}{}_{\kappa_3} \dots R^{r_n}{}_{r_1}{}^{\kappa_n}{}_{\kappa_1} . \tag{10}$$

This vanishes for hyperkähler metrics for all n. It is now almost trivial to show that $\nabla_{\hat{\imath}}\nabla_{\hat{\jmath}}S^{(n)}$ is Riemannian, and hence $T^{(n)}_{ij}$ is a candidate (n + 1)-loop counterterm, with $T^{(n)}_{ij}$ given by[2]

$$\begin{aligned} T^{(n)}_{ij} &\equiv \nabla_i \nabla_j S^{(n)} + \nabla_{\hat{\imath}} \nabla_{\hat{\jmath}} S^{(n)} \\ &= \nabla_i \nabla_j S^{(n)} + 2^{n-1} \Big\{ \nabla_{r_2} \nabla_{\kappa_2} R_j{}^{r_1}{}_i{}^{\kappa_1} + \nabla_{r_2} \nabla^{\kappa_1} R_j{}^{r_1}{}_{i\kappa_2} + [\nabla_i, \nabla_{r_2}] R_{j\kappa_2}{}^{r_1\kappa_1} \\ &\qquad + [\nabla_i, \nabla_{r_2}] R_j{}^{\kappa_1}{}_{\kappa_2}{}^{r_1} \Big\} R^{r_2}{}_{r_3}{}^{\kappa_2}{}_{\kappa_3} \dots R^{r_n}{}_{r_1}{}^{\kappa_n}{}_{\kappa_1} \\ &\quad + 2^{n-1} n \nabla_{r_2} R_j{}^{r_1\kappa_1}{}_{\kappa_2} \Big\{ (\nabla_{\kappa_3} R^{r_2}{}_{r_3 i}{}^{\kappa_2} + \nabla^{\kappa_2} \nabla^{r_2}{}_{r_3 i\kappa_3}) R^{r_3}{}_{r_4}{}^{\kappa_3}{}_{\kappa_4} \dots R^{r_n}{}_{r_1}{}^{\kappa_n}{}_{\kappa_1} \\ &\qquad + \dots \\ &\qquad + R^{r_2}{}_{r_3}{}^{\kappa_2}{}_{\kappa_3} R^{r_3}{}_{r_4}{}^{\kappa_3}{}_{\kappa_4} \dots (\nabla_{\kappa_1} R^{r_n}{}_{r_1 i}{}^{\kappa_n} + \nabla^{\kappa_1} R^{r_n}{}_{r_1 i\kappa_1}) \Big\} \end{aligned} \tag{11}$$

The constraints on a function S if it is to have the property that $\nabla_{\hat{\imath}}\nabla_{\hat{\jmath}}S$ is independent of J_{ij} are very restrictive, and it is tempting to conjecture that the examples of Eq. (10) are essentially exhaustive. Unfortunately there seems to be no obvious way of constructing the most general such function, and so in the light of past experience with σ-models, such speculations are probably ill-advised. What can be said is that beyond one loop, the counterterms must certainly take the cohomologically trivial form $\partial_a \partial_{\bar{b}} S$ on a Kähler space, where S is globally defined. Thus for a Kähler metric to yield a zero of the β-function, it must satisfy

$$R_{a\bar{b}} = \partial_a \partial_{\bar{b}} S \tag{12}$$

for some globally defined scalar formed from Riemann tensors and covariant derivatives. It follows therefore that M must have vanishing first Chern class. Whether Eq. (12) would admit non-trivial (i.e. non-hyperkähler) solutions is not clear. One can invoke Yau's theorem[9] to construct an infinite sequence of metrics $g^{(r)}_{ij}$, defined by

$$R_{a\bar{b}}(g^{(r+1)}) = \partial_a \partial_{\bar{b}} S(g^{(r)}) , \tag{13}$$

with $g_{ij}^{(o)}$ Ricci-flat, but making any rigorous statements concerning the convergence of the sequence would appear to be problematical [see, for example, Ref.10)].

3. Open strings in background gauge fields

We now turn to an example of a σ-model in which it appears to be possible to compute a β-function to all loop orders. It corresponds to the open bosonic string, propagating in a flat (26-dimensional) background, but in the presence of a non-zero gauge potential A_μ. This couples to the string only at its ends, and the action is

$$I = \int_\Sigma d^2\sigma\, \partial_+ X^\mu \partial_- X^\nu \eta_{\mu\nu} + \int_{\partial\Sigma} d\tau\, A_\mu \dot{X}^\mu . \tag{14}$$

Expanding around a classical solution X^μ, we write $X^\mu = \bar{X}^\mu + \Pi^\mu(\xi)$, where as usual, we employ a normal co-ordinate expansion in the quantum field ξ^μ around the point $\bar{X}^\mu$. Dropping the bar, we have

$$A_\mu \dot{X}^\mu \rightarrow A_\mu \dot{X}^\mu + F_{\mu\nu} \xi^\nu \dot{X}^\mu + \tfrac{1}{2} \nabla_\lambda F_{\mu\nu} \dot{X}^\mu \xi^\lambda \xi^\nu + F_{\mu\nu} \dot{\xi}^\mu \xi^\nu$$
$$+ \tfrac{1}{2} \nabla_\rho \nabla_\lambda F_{\mu\nu} \dot{X}^\mu \xi^\rho \xi^\lambda \xi^\nu + \tfrac{1}{3} \nabla_\lambda F_{\mu\nu} \dot{\xi}^\mu \xi^\lambda \xi^\nu + \cdots . \tag{15}$$

To compute the β-function, we must consider all one-particle irreducible diagrams that contribute counterterms to $\int A_\mu \dot{X}^\mu$ with $1/\varepsilon$ poles in dimensional regularization. This has been discussed at the one-loop level[11)], and extended to all loops in Ref. 12).

At one-loop, counterterms come from diagrams with one vertex $\nabla F\, X\, \xi\xi$ together with any number of insertions of the vertex $F\xi\xi$. It follows from symmetry arguments that only terms with an even number of such insertions contribute, and so schematically one has

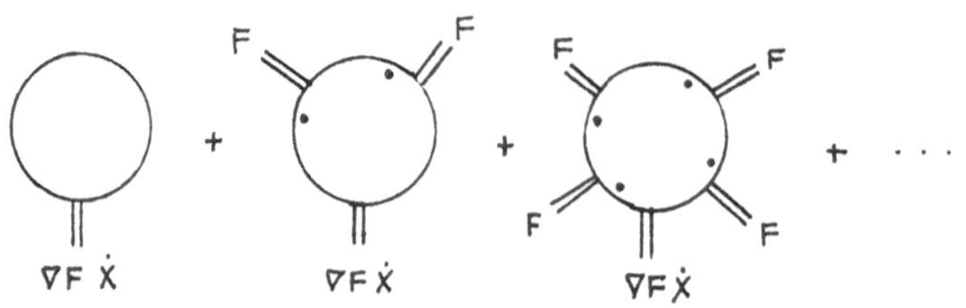

(16)

where the dots indicate quantum lines corresponding to $\dot{\xi}^\mu$. On the boundary $\partial\Sigma$, the propagator $G_{\mu\nu}$ takes the form $\log|\tau-\tau'|$, or in momentum space, $\tilde{G}_{\mu\nu} \sim \delta_{\mu\nu}/|p|$. Thus for all lines with dots, the time derivative on $\dot{\xi}^\mu$ brings down a p which cancels the propagator, yielding a δ-function $\delta(\tau-\tau')$. The result is that Eq. (16) reduces to

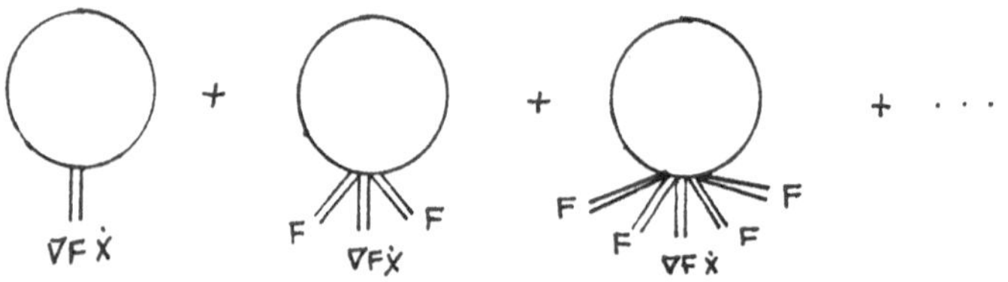

(17)

where the closed loop yields a factor $\int_{-\infty}^{\infty} dp/|p|$ common to all terms.

Introducing an infra-red mass regulator m, and dimensionally regularizing, this gives a factor

$$\int_0^\infty d^{(1+\varepsilon)}p \, (p^2+m^2)^{-1/2} \sim \frac{1}{\varepsilon} + 2(\gamma - \log m^2) + O(\varepsilon)$$

$$= \frac{1}{\varepsilon} + \text{finite} . \tag{18}$$

The infra-red divergence is irrelevant as far as the β-function is concerned, and so we obtain, taking account of combinatoric factors[11),]

$$\beta_\mu = \nabla_\lambda F_{\mu\nu} \left[1 + F^2 + F^4 + \cdots \right]^{\nu\lambda} , \tag{19}$$

which can be summed to give[11)]

$$\beta_\mu = \nabla_\lambda F_{\mu\nu} \left[(1-F^2)^{-1} \right]^{\nu\lambda} . \tag{20}$$

Here F^2, F^4, etc. denote matrices, with components $F_{\mu\rho}F^\rho{}_\nu$, $F_\mu{}^\rho F_\rho{}^\sigma F_\sigma{}^\lambda F_{\lambda\nu}$, etc. It was shown in Ref. 11) that although β_μ itself cannot be obtained from the variation of any action, the equation $\beta_\mu = 0$ is equivalent to that obtained from the Born-Infeld action

$$S = \int d^{26}X\, L \quad , \quad L = \sqrt{\det(\delta_{\mu\nu} + F_{\mu\nu})} \ . \tag{21}$$

This yields the field equation $W_\mu = 0$, where $\delta S \equiv \int L W^\mu \delta A_\mu$ and

$$W_\mu = [(1-F^2)^{-1}]_{\mu\nu} \nabla_\lambda F_{\nu\sigma} [(1-F^2)^{-1}]^{\sigma\lambda} . \tag{22}$$

Thus, from Eq. (20),

$$W_\mu = [(1-F^2)^{-1}]_{\mu\nu} \beta^\nu . \tag{23}$$

The matrix F^2 is negative semi-definite, and so $(1-F^2)$ is a non-degenerate matrix, implying that β_μ vanishes if and only if $F_{\mu\nu}$ is a solution of the Born-Infeld field equation (22)[11].

The extension of the β-function calculation to all loop orders[12] is most easily discussed by considering a typical diagram. From Eq. (15), one of the contributions at two loops will be given by

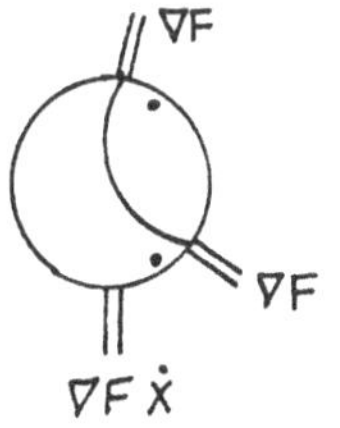

(24)

As before, the propagators on lines with dots are cancelled, giving

(25)

where each loop gives a factor $\int dp/p$. We again regularize using Eq. (18), but now we must take care to subtract out the one-loop subdivergent diagrams, by introducing appropriate counterterms. The net effect is that the remaining contribution from Eq. (25) is a purely $1/\varepsilon^2$ divergence, and thus does not affect the β-function, which is determined entirely by the $1/\varepsilon$ pole. By similar arguments one can show that at n-loops the counterterms contribute pure $1/\varepsilon^n$ poles, and again do not contribute to β_μ. There are two special cases that must be considered, in which dots become "trapped" in closed loops. This can happen with either one or two dots, giving factors of $\int dp$ or $\int p\,dp$ respectively. Both of these vanish in dimensional regularization.

We see therefore that the all orders expression for β_μ is precisely the same as the one-loop expression Eq. (20)[12]. Of course this result is highly dependent on the renormalization scheme used. It is consistent with the results of a three-loop calculation in Ref. 13).

In view of the conjectured equivalence of demanding conformal invariance of the σ-model on the one hand, and solutions of the string equations of motion on the other, the above result appears somewhat surprising at first sight. String scattering calculations yield amplitudes with non-polynomial dependence on the external momenta, and thus the string effective action will involve arbitrarily many derivatives on $F_{\mu\nu}$, whereas Eq. (21) involves none at all. However, there need not necessarily be any conflict between these results, since it is solutions of the two systems of equations of motion that are supposed to be in correspondence, which does not necessarily mean that the equations themselves must have the same form.

Finally, we remark that it was also shown in Ref. 12) that the same Born-Infeld action, Eq. (21), arises in the case of the ten-dimensional superstring in an arbitrary purely bosonic gauge field background. This result[12], obtained for the Neveu-Schwarz-Ramond action, is different from that obtained by Tseytlin[14], who used the Green-Schwarz action. It is not clear what the origin of this discrepancy is; it may perhaps be an indication that in a non-trivial background, the two formalisms are not equivalent. This may be similar to the situation in the case of closed superstrings in curved background spacetimes[15),10)].

ACKNOWLEDGEMENTS

I am grateful to E. Bergshoeff, M. Freeman, E. Sezgin, M. Sohnius, K. Stelle and P. Townsend, with whom the work described in this paper was carried out, for extensive discussions.

REFERENCES

1) D.J. Gross and E. Witten, Nucl. Phys. B277 (1986), 1.

2) M.D. Freeman, C.N. Pope, M.F. Sohnius and K.S. Stelle, Phys. Lett. 178B (1986), 199.

3) M.T. Grisaru, A.E.M. Van de Ven and D. Zanon, Phys. Lett. 173B (1986), 423; Nucl. Phys. B277 (1986), 388 and 409.

4) M.D. Freeman and C.N. Pope, Phys. Lett. 174B (1986), 48.

5) M.T. Grisaru and D. Zanon, Phys. Lett. 177B (1986), 347.

6) C.M. Hull, Nucl. Phys. B260 (1985), 182;
L. Alvarez-Gaumé and P. Ginsparg, Comm. Math. Phys. 102 (1985), 311.

7) P.S. Howe, G. Papadopoulos and K.S. Stelle, Phys. Lett. 174B (1986), 405

8) C.N. Pope, M.F. Sohnius and K.S. Stelle, Nucl. Phys. B283 (1987), 192.

9) S.T. Yau, Proc. Natl. Acad. Sci. 74 (1977), 1798.

10) M.D. Freeman, C.N. Pope, M.F. Sohnius and K.S. Stelle, CERN preprint TH.4632/87, to appear in the Proceedings of the Colloquium on Strings and Gravity, Meudon, Paris, September 1986.

11) A. Abouelsaood, C.G. Callan, C.R. Nappi and S.A. Yost, Princeton preprint (1986).

12) E. Bergshoeff, C.N.Pope, E. Sezgin and P.K. Townsend, Trieste preprint (1987), to appear in Phys. Lett. B.

13) H. Dorn and H.J. Otto, Z. Phys. C 32 (1986), 599.

14) A.A. Tseytlin, Nucl. Phys. B273 (1986), 391.

15) M.D. Freeman, C.M. Hull, C.N. Pope and K.S. Stelle, Phys. lett. 185B (1987), 351.

THE d=2 CONFORMALLY INVARIANT SU(2) σ-Model WITH WESS-ZUMINO TERM AND RELATED CRITICAL THEORIES[+)]

R. Flume
Physikalisches Institut, Universität Bonn
Nussallee 12, D-5300 Bonn 1

Abstract: The structure of the four point correlation functions of the conformally invariant SU(2) σ-model is presented. Relations of the SU(2) model to other critical theories are pointed out.

I.) After the poineering work of Belavin, Polyakov and Zamolodchikov /1/ much progress has been made during the last two years in the analysis of two-dimensional critical systems in statistical mechanics and field theory /2/. The list of field theoretical models which so far have been solved [f1)] and are partially identified with critical systems of statistical mechanics contains

i) all unitary models with central Virasoro charge $c < 1$ /3/-/6/
ii) unitary theories with N=1 and N=2 supersymmetry ($c < \frac{3}{2}$ (N=1), $c < 3$ (N=2)) /7/ - /10/
iii) models realising parafermion algebras /11/
iv) σ-models on group manifolds with Wess-Zumino term /12/ - /17/.

I want to report in this contribution on recent work specially devoted to the SU(2) model /14/ - /16/ and to point out links of this model with the other theories (i) - (iii) in the above given list. (Technical - not conceptual - complications have till now impeded the explicit solution - that is, the construction of correlation functions - of σ-models on groups of higher rank).

II) The observation that σ-models on group manifolds aquire through the addition of a Wess-Zumino term a conformally invariant fixed point at a non-zero value of the coupling constant is due to Witten /12/. I will not discuss the Lagrangian version of the "Wess-Zumino-Witten (WZW) model". Instead I assume following Knizhnik and

---

+) Talk presented at the Ringberg Workshop "Renormalization of Quantum Field Theories with non-linear Field Transformations", Feb. 16 - 20, 1987.

f1) At least in the sense that the structure of the operator product algebra is known.

Zamolodchikov /13/ - the existence of two Kac-Moody algebras of currents as constitutive starting point. Let G be a simple compact group. We suppose that in the theory under consideration two sets of currents $\{j^a_L\}$ and $\{j^a_R\}$ occur generating the symmetry group $G_L \otimes G_R$ ($G_R = G_L = G$) (a=1, ..., dim G labels a basis of the Lie-algebra of G). The general frame work is two-dimensional Euclidean field theory. Let z (with complex conjugate $\bar{z}$) be a complex coordinate for the Euclidean plane. The currents j^a_L and j^a_R are supposed to satisfy the conservation equation

$$\partial_{\bar{z}}\, j^a_L = 0 = \partial_z\, j^a_L \quad ,$$

and to obey the commutation relations - conveniently quoted in terms of a Laurent decomposition $j^a_{L,n} = \frac{1}{2\pi i}\oint d\xi\, \xi^n j^a_L(\xi)$, $j^a_{R,n} = \frac{1}{2\pi i}\oint d\bar{\xi}\, \bar{\xi}^n j^a_R(\bar{\xi})$ –

$$\left[j^a_{R,n} , j^b_{R,m} \right] = f^{abc}\, j^c_{R,n+m} - \frac{k}{2}\, n\, \delta^{ab}\, \delta_{n+m,0} \qquad (1a)$$

$$\left[j^a_{L,n} , j^b_{L,m} \right] = f^{abc}\, j^c_{L,n+m} - \frac{k}{2}\, n\, \delta^{ab}\, \delta_{n+m,0} \qquad (1b)$$

$$\left[j^a_{L,n} , j^b_{R,m} \right] = 0 \qquad (1c)$$

f^{abc} denote here totally antisymmetric structure constants of the Lie algebra of G. The normalization is assumed to be chosen such that the long roots of the Lie algebra have unit length. The parameter k on the r.h.s. of (1a) and (1b), the so called Kac-Moody central charge, is taken to be a positive integer number. (This is the necessary and sufficient condition to ensure the existence of unitary highest weight representations of the Kac-Moody algebra (1a) and (1b), cf. /18/, /19/). The energy momentum tensor in the field theoretical sector generated by the currents is known to be of the Sugawara form. Its two independent components taken into the z-z and $\bar{z}$-$\bar{z}$ directions read as (/20/)

$$T_{zz} = - \frac{1}{k + c_V} \sum_a : j^a_L\, j^a_L : \qquad (2a)$$

$$T_{\bar{z}\bar{z}} = - \frac{1}{k + c_V} \sum_a : j^a_R\, j^a_R : \qquad (2b)$$

where the double points denote normal ordering with respect to current frequencies. C_V is the eigenvalue of the quadratic Casimir operator in the adjoint representation. Energy momentum conservationis expressed through

$$\partial_{\bar{z}} T_{zz} = 0 = \partial_z T_{\bar{z}\bar{z}} \quad .$$

The Laurent components of T_{zz} and $T_{\bar{z}\bar{z}}$

$$L_n = \frac{1}{2\pi i} \oint d\xi \, T_{zz}(\xi) \, \xi^{n+1} \, , \, \bar{L}_n = \frac{1}{2\pi i} \oint d\bar{\xi} \, T_{\bar{z}\bar{z}}(\bar{\xi}) \, \bar{\xi}^{n+1}$$

generate the Virasoro algebras

$$[L_n , L_m] = (n-m) L_{n+m} + \frac{c}{12} (n^3 - n) \delta_{n+m,0} \tag{3a}$$

$$[\bar{L}_n , \bar{L}_m] = (n-m) \bar{L}_{n+m} + \frac{c}{12} (n^3 - n) \delta_{n+m,0} \tag{3b}$$

$$[L_n , \bar{L}_m] = 0 \tag{3c}$$

$$c = \frac{k \, d_G}{k + C_V} \quad , \quad (d_G = \dim G) \tag{3d}$$

Eqs. (3a) - (3d) are straightforward consequences of the Kac-Moody algebra relations (1a) - (1c).

The Sugawara form of the energy momentum tensor supplies the key to the solution of the WZW models. The observation is due to Knizhnik and Zamolodchikov /13/(cf. also Dashen and Frishman, ref. /20/).To explain their observation we have to introduce the notion of primary fields /1/: An operator φ is called primary with respect to the Kac-Moody and Virasoro algebras (1) and (3) if the following highest weight relations are satisfied,

$$L_n \varphi = \bar{L}_n \varphi = j^a_{R,n} \varphi = j^a_{L,n} \varphi = 0 \qquad \forall n > 0 \quad ,$$

$$L_0 \varphi = \Delta_\varphi \varphi \, , \, \bar{L}_0 \varphi = \bar{\Delta}_\varphi \varphi \, , \, j^a_{R,0} \varphi = t^a_{R,\varphi} \varphi \, , \, j^a_{L,0} \varphi = t^a_{L,\varphi} \varphi \, .$$

Δ_φ and $\bar{\Delta}_\varphi$ are here the scaling dimensions of φ with respect to dilations in z and $\bar{z}$ resp.. $t^a_{L,}\varphi$ and $t^a_{R,}\varphi$ denote representation matrices of the Lie algebra of G_R and G_L resp. (φ is supposed to stand for an operator multiplet carrying some irreducible representations of the two groups). $J^a_{L,n}\varphi$, $L_n\varphi$ etc. are shorthands for

$$j^a_{L,n}\,\varphi(z) = \frac{1}{2\pi i}\oint d\xi\,(\xi - z)^n\, j^a_L(\xi)\,\varphi(z)$$

$$L_n\,\varphi(z) = \frac{1}{2\pi i}\oint d\xi\,(\xi - z)^{n+1}\, T_{zz}(\xi)\,\varphi(z)$$

All non-primary operators of the theory are determined through the primary ones via representation theory of the two algebras (1) and (3)[f2]. The solution of the current and energy-momentum Ward identities of correlation functions

$$< \varphi_1(z_1,\bar z_1)\ .\ .\ .\ \varphi_n(z_n,\bar z_n) >$$

of primary operators φ_i (with scaling dimensions Δ_i, $\bar\Delta_i$ and carrying representations $\{t_{L,i}\}$ and $\{t_{R,i}\}$ resp.) are particularly simple. One finds

$$< j^a_R(z)\,\varphi_1(z_1)\,..\,\varphi_n(z_n)> = \sum_{i=1}^{n}\frac{t^a_{R,i}}{z - z_i}\,<\varphi_1(z_1)\,..\,\varphi_n(z_n)> \tag{4}$$

$$< j^a_L(\bar z)\,\varphi_1(\bar z_1)\,..\,\varphi_n(\bar z_n)> = \sum_{i=1}^{n}\frac{t^a_{L,i}}{\bar z - \bar z_i}\,<\varphi_1(\bar z_1)\,..\,\varphi_n(\bar z_n)> \tag{5}$$

$$\begin{aligned}< T_{zz}(z)\,\varphi_1(z_1)\,..\,\varphi_n(z_n)> &= \\ &= \sum_{i=1}^{n}\left(\frac{\Delta_i}{(z - z_i)^2} + \frac{1}{z - z_i}\partial_{z_i}\right)<\varphi_1(z_1)\,..\,\varphi_n(z_n)>\end{aligned} \tag{6}$$

$$\begin{aligned}< T_{\bar z\bar z}(\bar z)\,\varphi_1(\bar z_1)\,..\,\varphi_n(\bar z_n)> &= \\ &= \sum_{i=1}^{n}\left(\frac{\bar\Delta_i}{(\bar z - \bar z_i)^2} + \frac{1}{\bar z - \bar z_i}\partial_{\bar z_i}\right)<\varphi_1(\bar z_1)\,..\,\varphi_n(\bar z_n)>\end{aligned} \tag{7}$$

Knizhnik and Zamolodchikov compare the Ward identities (6) and (7) with the following relations derived from the operator product expansion.

---

f2) It is assumed here tacitely that the identity is the unique operator in the theory which commutes with all currents j^a_L and j^a_R .

$$T_{zz}(z)\,\varphi_i(z_i) = \frac{\frac{c_{L,i}}{k+c_v}}{(z-z_i)^2}\,\varphi_i(z_i) - \left(\frac{2}{k+c_v}\right)\frac{1}{(z-z_i)} : J_L^a\, t_{L,i}^a\, \varphi_i :(z_i) + \tag{8}$$

$$+ \text{ regular terms}$$

$$T_{\bar{z}\bar{z}}(\bar{z})\,\varphi_i(\bar{z}_i) = \frac{\frac{c_{R,i}}{k+c_v}}{(\bar{z}-\bar{z}_i)^2}\,\varphi_i(\bar{z}_i) - \left(\frac{2}{k+c_v}\right)\frac{1}{(\bar{z}-\bar{z}_i)} : J_R^a\, t_{R,i}^a\, \varphi_i :(\bar{z}_i) + \tag{9}$$

$$+ \text{ regular terms}$$

where $c_{L,i}$ ($c_{R,i}$) denotes the eigenvalue of the quadratic, Casimir operator, in representation carried by φ_i. Identifying the singular terms in $(z-z_i)$ of Eqs. (6) and (8) ((7) and (9)) one obtains first for the scaling dimensions the relations

$$\Delta_i = \frac{c_{L,i}}{k+c_v} \quad , \quad \bar{\Delta}_i = \frac{c_{R,i}}{k+c_v}$$

and second, after use of the current Ward identities (5) ((6)) a system of differential equations - the "Knizhnik-Zamolodchikov equations" - for the correlation functions of primary fields

$$\partial_{z_i} \langle \varphi_1(z_1) \ldots \varphi_n(z_n) \rangle =$$

$$= -\frac{1}{\kappa} \sum_{\substack{j,a \\ j \neq i}} \frac{t_{L,i}^a \otimes t_{L,j}^a}{z_i - z_j} \langle \varphi_1(z_1) \ldots \varphi_n(z_n) \rangle \tag{10}$$

$$\kappa = \frac{1}{2}(k + c_v)$$

and an analogous system of differential equations in the complex conjugate variables $\bar{z}_i$.

The general solution of the Knizhnik-Zamolodchikov equations for the four point functions of the SU(2) σ-model has been found in references /14/ and /15/. Let $\{\varphi m^{(z)}_{(j)}\} = \varphi_j(z)$ be a primary operator multiplet carrying the isospin j representation of $G_L = SU(2)$[f3]. I choose isospins $j_i \ldots j_4$ satisfying the relations

$$j_1 \leq j_2 \leq j_3 \leq j_4 = \sum_{i=1}^{3} j_i - \ell \,, \quad \ell \leq 2 j_1 \,, \quad \ell \in \mathbb{N} \tag{11}$$

f3) I ignore for the moment the $\bar{z}$-dependence of the operators and do also not specify under what representation φ_j transforms with respect to the right handed SU(2).

Global SU(2) invariance implies that in a correlation

$$\langle \varphi_{j_1}(z_1)\dots\varphi_{j_4}(z_4)\rangle \equiv F_{j_1,\dots,j_4}(z_1,\dots,z_4)$$

the isospins $j_i \dots j_4$ are combined into SU(2) singlets. With the special choice ((11)) of isospins one can form (l+1) different singlets and to this corresponds a decomposition of $F_{j_1 \dots j_4}$ into (l+1) components:

$$F_{j_1,\dots,j_4} = \begin{pmatrix} F^{(1)}_{j_1,\dots,j_4} \\ \vdots \\ F^{(\ell+1)}_{j_1,\dots,j_4} \end{pmatrix}$$

Ths basis of invariants I choose is constructed by projection of the tensor product of representations of isospin j_1 and j_2 into the representation with isospin $j_1 + j_2 - (k-1)$, $k = 1, \dots, l+1$ and contraction with the same representation combining j_3 and j_4. The Knizhnik-Zamolodchikov equation becomes an (l+1)-dimensional matrix differential equation for the vector function F. Its general solution can be represented in terms of a l-dimensional contour integral of the Euler type (/14/, /15/)

$$\begin{aligned} F^{(k)} = & \prod_{i<s<4} (z_i - z_s)^{\frac{j_i j_s}{\kappa}} \prod_{i<4} (z_i - z_4)^{-\frac{j_i(j_4+1)}{\kappa}} \cdot \\ & \cdot (z_1 - z_2)^{k-1} (z_3 - z_4)^{\ell+1-k} \int\dots\int \prod_{i=1}^{\ell} dt_i \prod_{i<j} (t_i - t_j)^{\frac{1}{\kappa}} \cdot \\ & \cdot \prod_{i=1}^{k-1} (t_i - z_1)^{-\frac{j_1}{\kappa}-1} (t_i - z_2)^{-\frac{j_2}{\kappa}-1} (t_i - z_3)^{-\frac{j_3}{\kappa}} (t_i - z_4)^{\frac{j_4+1}{\kappa}} \cdot \\ & \cdot \prod_{i=k}^{\ell} (t_i - z_1)^{-\frac{j_1}{\kappa}} (t_i - z_2)^{-\frac{j_2}{\kappa}} (t_i - z_3)^{-\frac{j_3}{\kappa}-1} (t_i - z_4)^{\frac{j_4+1}{\kappa}-1} \end{aligned} \qquad (12)$$

Several comments and remarks are to be made.

i) The sum of exponents attached to any of the integration variables $t_1 \dots t_l$ is equal to -2. The solution (12) could so be interpreted as integral over vertex operators of scaling dimension 1 (with respect to scaling in z) if the interaction of these "vertices" with the external points $z_1 \dots z_4$ were not slightly unsymmetric.

ii) The matrix differential equation has in the case under consideration a fundamental system of (l+1) independent vector solutions. Those are found by making different choices of contours in (12). In order to construct a proper four point function one has to consider in addition the general solution of the Knizhnik-Zamolodchikov equations in the complex conjugate variables $\bar{z}_i$ and then to superimpose products of special solutions of the z- and $\bar{z}$-equations so that the resulting expression is a one-valued function in the Euclidean plane. This program is known as the conformal bootstrap. It fixes in particular up to an overall normalization the expansion coefficients of the operator algebra. For details of the conformal bootstrap I refer to /14/ and /15/.

iii) The functions given by Eq. (12) are identical in structure with the four point functions of certain closed subalgebras of statistical mechanics systems with central charge $c < 1$. (cf. /1/). The ties of the SU(2) model with these and other systems will be discussed in the following section.

III) A general recipe to construct the correlation functions of critical systems with $c < 1$ has been given by Dotsenko and Fateev /6/. The four point functions found by these authors have the general form

$$F_{m,n}(z_1,..,z_4) = \prod_{i<j}(z_i - z_j)^{\gamma_{ij}} \cdot \int..\int \prod_{i=1}^{m} dt_i \prod_{j=1}^{n} dt'_j \prod_{i,j}(t_i - t'_j)^{-2} \cdot f_m(t_1,..,t_m; z_1,..,z_4)\, \hat{f}_n(t'_1,..,t'_n; z_1,..,z_4) \tag{13}$$

$$f_m(t_1,..,t_m; z_1,..,z_4) = \prod_{i=1}^{m}\prod_{s=1}^{4}(t_i - z_s)^{a_s} \prod_{i<j}(t_i - t_j)^{b}$$

$$\hat{f}_n(t'_1,..,t'_n; z_1,..,z_4) = \prod_{i=1}^{n}\prod_{s=1}^{4}(t'_i - z_s)^{\hat{a}_s} \prod_{i<j}(t'_i - t'_j)^{\hat{b}}$$

where the exponents a_s ($\hat{a}_s$), b ($\hat{b}$) are functions of c and m (n). Comparing with Eq. (12) one sees that for the case $m = 0$ or $n = 0$ in Eq. (13) one encounters the same type of contour integrals as in the SU(2) model. Those special cases correspond to closed operator subalgebras of the $c < 1$ models.

The functions $F_{n,m}$ represent the general solution of a linear differential equation of $(n+1)\cdot(m+1)$-th order which can be derived from degeneracy relations of the Virasoro algebra. Special solutions are again given through specific choices of contours. To be able to execute the conformal bootstrap program, as it has been sketched in the previous section, one has to know how different special solutions are related to each other under analytic continuation. One needs in other words information on the monodromy group of the system of functions represented by the contour integral (13). The problem is facilated through the observation of Dotsenko and Fateev (6) that the monodromy group factorizes on the pieces f_m and $\hat{f}_n$ in the integrand on the r.h.s. of Eq. (13). It means, that the monodromy groups of $F_{o,n}$ and $F_{m,o}$ determine completely the monodromy of $F_{m,n}$. One is so in this respect lead to the considera-tion of two seperate SU(2)-like situations. The analysis of closed operator algebras in the $c < 1$ systems can in fact be deduced step by step from the corresponding analysis in the SU(2) model (/16/, /21/).

Another interesting connection between statistical systems with $c < 1$ and the SU(2) model has been observed by Gepner /22/. He notes that the characters of the highest weight SU(2) Kac-Moody representations transform under the modular group very similiarly as the characters of the highest weight Virasoro representations in the $c < 1$ systems. Gepner uses the similiarity to relate modular invariant combinations of SU(2) Kac-Moody characters to modular invariant partition functions of the statistical systems. The subject has further been persued by Cappelli, Itzykson and Zuber /23/.

Still another aspect relating the representation theory of the SU(2) Kac-Moody algebra and the Virasoro algebra with $c < 1$ has been pointed out by Goddard, Kent and Olive /24/: Start with a theory realising the tensor product of two SU(2) Kac-Moody algebras (that is, $G_L = G_R = SU(2) \otimes SU(2)$) with central charges k_1 and k_2 for the two factors. Suppose that one can construct a sector in which the diagonal subgroup of $SU(2) \otimes SU(2)$ is annihilated. In this sector are realised, if k_1 and k_2 are properly chosen, the highest weight representations of some $c < 1$ Virasoro algebra. Goddard, Kent and Olive show that in this way all representations of all $c < 1$ unitary models can be reached. There should be a conceptual link between the algebraic Goddard-Kent-Olive procedure and the above mentioned field theoretical relations which has to be found.

The Dotsenko-Fateev Coulomb gas construction is easily supersymmetrized by adding a free Majorana field. The unitary theories with N=1 supersymmetry have been worked out along these lines in ref. /8/. The previous remark on monodromy groups applies also in this case.

I finally mention the works of Fateev and Zamolodchikov /11/ and of Qiu /10/. It was shown in ref. /11/ that theories with an underlying parafermion algebra can be mapped isomorphically on the SU(2) WZW model. Qiu /10/ found recently the operator solution of the degenerate critical systems with N=2 supersymmetry in terms of operatorsof parafermion theories. The N=2 theories are therefore also by the isomorphism of Fateev and Zamolodchikov related to the SU(2) WZW model.

References

/ 1/ A.A. Belavin, A.M. Polyakov and A.B. Zamolodchikov, Nucl.Phys. B241 (1984) 333.

/ 2/ J.L. Cardy, in phase Transitions and Critical Phenomena vol. 11, ed. C.Domb and J.L. Lebowitz (Academic Press: New York, 1986), and references therein.

/ 3/ D. Friedan, Z. Qiu and S.H. Shenker, in Vertex Operators in Mathematics and Physics, eds. J. Lepowsky et al. (Springer 1984); Phys. Rev. Lett. 52 (1984) 1575.

/ 4/ G.F. Andrews, R.J. Baxter and P.I. Forrester, J. Stat. Phys. 35 (1984) 193.

/ 5/ D.A. Huse, Phys. Rev. B 30 (1984) 3908.

/ 6/ V.S. Dotsenko and F.A. Fateev, Nucl. Phys. B240 (1984) 312.

/ 7/ D. Friedan, Z. Qiu and S. Shenker, Phys. Rev. Lett. 52 (1984) 1575.

/ 8/ M. Bershadsky, V. Knizhnik and M Teitelman, Phys. Lett.151b (1985) 31.

/ 9/ W. Boucher, D. Friedan and A. Kent, Phys. Lett. B172(1986) 316;
P. di Vecchia, J.L. Petersen and H.B. Zheng, Phys. Lett. 174B (1986) 280;
S. Nam, Phys. Lett. B172 (1986) 323.

/10/ Z. Qiu, Institute for Advanced Study, Princeton preprint (October, 1986).

/11/ A.B. Zamolodchikov and V.A. Fateev, Zh. Eksp. Teor. Fiz 89 (1985) 380, (Sov. Phys. JETP 62, 215).

/12/ E. Witten, Comm. Math. Phys. 92 (1984) 455.

/13/ V.G. Knizhnik and A.B. Zamolodchikov, Nucl. Phys. B247 (1984) 83.

/14/ V.A. Fateev and A.B. Zamolodchikov, Yad. Fiz. 43 (1986) 75.

/15/ P. Christe and R. Flume, Nucl. Phys. B282 (1987) 466.

/16/ P. Christe and R. Flume, Phys. Lett. B (1987) in press.

/17/ D. Gepner and E. Witten, Princeton University preprint (April 1986).

/18/ V.G. Kaç, Infinite - dimensional Lie algebra - an Introduction (Birkhäuser, 1983).

/19/ P. Goddard and D. Olive, in Vertex Operators in Mathematics and Physics, eds. J. Lepowsky et al. (Springer 1984).

/20/ R. Dashen and Y. Frishman, Phys.Rev. D11 (1975) 2781.

/21/ P. Christe, work in preparation.

/22/ D. Gepner, Princeton University preprint (1986).

/23/ A. Cappelli, C. Itzykson and J.-B. Zuber, Saclay preprint PhT 86/122 (1986).

/24/ P. Goddard, A. Kent and D. Olive, Comm. Math. Phys. 103 (1986) 195.

THE TWO-DIMENSIONAL O(n) NONLINEAR σ-MODEL FROM A WILSON RENORMALIZATION GROUP VIEWPOINT

P.K. MITTER and T.R. RAMADAS
Laboratoire de Physique Théorique et Hautes Energies*
Université Pierre et Marie Curie, Paris

* Laboratoire associé au CNRS UA 280

and

School of Mathematics, Tata Institute of Fundamental Research
Homi Bhabha Road, Bombay 400 005

1° INTRODUCTION

Consider the O(n) nonlinear σ-model in two (euclidean) dimensions. With a lattice cut off -on a lattice of spacing a- the theory is defined by R^n-valued, spins φ on a lattice with the constraint $|\varphi|^2 = \frac{1}{Z(a)}$ and the bare action

$$S_1(a) = \frac{Z(a)}{2g_0^2(a)} \cdot \frac{1}{2} \sum_{\substack{n.nhb \\ i,j}} | \varphi(i) - \varphi(j) |^2 \tag{1.1}$$

It is believed that (Polyakov [1], Brézin et al [2,3])

i) The theory has a unique phase for all valness of Z,g. This phase has exponentially decreasing correlation functions.

ii) A scaling limit of the theory exists as a continuum field theory with a mass-gap, provided that as a $\to$ 0 we let $g_0(a) \to 0$ and $Z(a) \to 0$ at rates computable in perturbation theory.

It is clearly important, to see what Wilson [4] renormalization group techniques or phase cell expansions (Glimm-Jaffe [5]) so successfully applied to renormalizable asymptotically free 4-fermionic theories [15,16] in 2-d (Gawedski-Kupiainen [6], Feldman et al [7]) can tell us about the 2-d non-linear σ-model. But here we immediately face the problem that the simplest and most appealing renormalization group transformations destroy the δ-function constraint in the model. Here we adopt the prescription : drop the constraint $|\varphi|^2 = \frac{1}{Z(a)}$ and take as bare action

$$S(a)= \frac{Z(a)}{2g_0^2(a)} \frac{1}{2} \sum_{\substack{n.nhb \\ i,j}} |\varphi(i)-\varphi(j)|^2 + \tilde{\lambda} \sum_i \left(Z|\varphi(i)|^2 - 1\right)^2 \qquad (1.2)$$

Then for large $\tilde{\lambda}$ we are approximating the σ-model. The idea is to let $\tilde{\lambda}$ go to infinity as $a \to 0$. Of course $\tilde{\lambda}$ has to go to infinity at some minimal rate, and this will be a crucial concern below.

We have three sets of results to report. The first (section 2) is a perturbative computation implementing the above idea. After applying a magnetic field to break the vacuum degeneracy and to get perturbation theory started, we show that up to one loop we can recover the usual perturbative results, including the renormalization constants provided $\tilde{\lambda}$ goes to infinity at a certain minimal rate. In fact this rate is given by :

$$\tilde{\lambda} = \frac{\lambda}{2g_0^2(a)} \qquad (1.3)$$

where λ is an arbitrary positive dimensionless constant and $g_0^2(a)$ is given by the 2-loop asymptotic freedom formula.

In Section 3 we try to understand the success of this perturbative computation by studying UV cutoff removal using Wilson RG transformations. We use Pauli-Villars regularization and a continuous Wilson RG which we study in a "local approximation" (this is the first step of a more complete study). We show that under RG flow (and up to $O(g^2)$) the parameter λ is driven to a fixed point. We show how asymptotic freedom is recovered, and show how to compute the renormalized trajectory in weak coupling in $g^2(M)$ the running coupling constant. Then g^2 turns out to be the only "marginal" variable, all other parameters contracting to fixed point values (there are no "relevant" parameters). This is completely consistent with the previous perturbative computation except that we miss out the wave function renormalization due to the local approximation.

Our third set of results (section 4) is from a study of the Wilson approximate recursion [Wilson [4]] for the model. This can also be interpreted as the renormalization group transformation for a certain hierarchical version of the model. This yields a fixed point in the variable $\tilde{\lambda}$ but not in λ . The theory can be renormalized around this fixed point, and would be asymptotically free. At the present however, we believe that this fixed point is an artifice of the hierarchical recursion we have chosen, since its existence in the full model would cause problems.

This work was begun in January 1986. In the meanwhile Gawedzki-Kupiainen have published (Gawedzki-Kupiainen [13]) a construction of the σ-model in a different hierarchical version. The picture we have presented in Section 3 is in qualitative agreement with theirs.

We thank Sourendu Gupta for help with numerical computations and Giovanni Felder for a discussion of his work on the local approximation to the Renormalization Group. P.K. Mitter thanks the Tata Institute of Fundamental Research and the National Board of Higher Mathematics (India) for hospitality. T.R. Ramadas thanks the LPTHE, Université Paris VI for hospitality.

2° THE PERTURBATION THEORY

Consider the action (with $\varphi = \tilde{\sigma} e_1 + \pi$, $\pi \cdot e_1 = 0$)

$$S_\Lambda = \frac{1}{2} \tilde{Z} \int d^2x \left(|\nabla \varphi|^2 + \mu_0^2 |\varphi|^2 \right) - \tilde{Z}^{1/2} \int d^2x \, \frac{\mu_0^2 \tilde{\sigma}}{g_0}$$
$$+ \frac{m_\sigma^2 g_0^2}{8} \int d^2x \left(\tilde{Z} |\varphi|^2 - \frac{1}{g_0^2} \right)^2 \tag{2.1}$$

Here Λ is a momentum cut-off and the parameters $\tilde{Z}$, μ_0^2, m_σ^2 and g_0 are Λ-dependent in a way to be determined. If we let $m_\sigma^2 \to \infty$ (for fixed Λ) this becomes the σ-model with a wave-function renormalization $Z = \tilde{Z} g_0^2$ and the addition of a magnetic field (the mass term $\mu_0^2 |\varphi|^2$ becomes in this limit a constant). We are, however, going to let $m_\sigma^2 \to \infty$ along with the cut-off. Note that, in comparing (1.2) and (2.1)

$$\tilde{\lambda} = \frac{m_\sigma^2 a^2}{8 g_0^2(a)} , \quad \Lambda \sim \frac{1}{a} \tag{2.2}$$

(In particular, $m_\sigma^2 = 4 \Lambda^2 \lambda$ in terms of (1.3)).
Define $\tilde{\sigma} = \sigma + \frac{1}{g_0 \tilde{Z}^{1/2}}$ we have then (with $\bar{m}_\sigma^2 = m_\sigma^2 + \mu_0^2$)

$$S_\Lambda(\;\; = \frac{1}{2} \tilde{Z} \int d^2x \left(|\nabla \pi|^2 + |\nabla \sigma|^2 + \mu_0^2 |\pi|^2 + \bar{m}_\sigma^2 |\sigma|^2 \right) \tag{2.3}$$
$$+ \int d^2x \left[\frac{m_\sigma^2 g_0^2}{8} \tilde{Z}^2 \left(|\pi|^2 + \sigma^2 \right)^2 + \frac{m_\sigma^2 g_0}{2} \tilde{Z}^{3/2} \left(|\pi|^2 + \sigma^2 \right) \sigma \right]$$

We can now study the perturbation theory in g_0. We summarise the (one-loop) computations in figures 1,2 and 3. (The computations were done with $\tilde{Z}=1$ and the

Green's functions then rescaled. Thus in the figures $\tilde{Z}=1$).

The computations are done under the assumption $m_\sigma^2/\Lambda^2 \gg 1$. We also assume $\mu_0^2/\Lambda^2 \ll 1$ (and this is consistent with the way μ_0 gets renormalized). The effect of the first assumption is to "contract the σ-lines" so that, for example

we obtain then (upto $O\left(\frac{\Lambda^2}{m_\sigma^2}\right)$), for the Fourier transform of the "pion" two point function $\langle \pi(x) . \pi(y) \rangle$, at 0-momentum

$$G^{(2)}_{\tilde{Z}=1}(0) = \frac{(n-1)}{\mu_0^2}\left\{1 - \frac{(n-1)}{2} g_0^2 \, I_{\Lambda,\mu_0}\right\} \qquad (2.3)$$

where
$$I_{\Lambda,\mu_0} = \int_{|k|\leq\Lambda} \frac{d^2k}{k^2+\mu_0^2}$$

and for the proper 1-pion irreducible four point function at 0-momentum

$$\Gamma^{(4)}_{\tilde{Z}=1}(0) = -\mu_0^2 (n-1)(n+1)\left[1 + \frac{3}{2}(n-1) I_{\Lambda,\mu_0}\, g_0^2\right] g_0^2 \qquad (2.4)$$

Note that the quadratically divergent diagrams have cancelled, leaving only logarithmic divergences.

If we take

$$\begin{aligned} g_0^2 &= Z_1 g^2 , \quad \mu_0^2 = \frac{Z_1^{1/2}}{\tilde{Z}^{1/2}} \mu^2 \\ Z_1 &= 1 - (n-2) g^2 I_{\Lambda,\mu} \\ \tilde{Z} &= 1 - g^2 I_{\Lambda,\mu} \\ (\Rightarrow Z &= 1 - (n-1) g^2 I_{\Lambda,\mu}) \end{aligned} \qquad (2.5)$$

we get finiteness of both $\Gamma^{(4)}_{ren.}(0) = \tilde{Z}^2 \Gamma^4_{\tilde{Z}=1}(0)$ and $\Gamma^{(2)}_{ren.}(0)$. This agrees with the computations of Brézin-Zinn Justin [2].

In fact the conditon $\frac{m_\sigma^2}{\Lambda^2} \gg 1$ is not necessary. As long as $\frac{m_\sigma^2}{\Lambda^2}$ is a positive constant we obtain the same physical results, with only finite changes in the renormalization constants.

3° THE WILSON RENORMALIZATION GROUP AND CUT OFF REMOVAL

Our starting point is the bare action :

$$S_0^{(\Lambda_N, N)}(\varphi) = \frac{1}{2}\int d^2x \left| e^{-\frac{\Delta}{2\Lambda_N^2}} \nabla\varphi \right|^2 + V_0^{(\Lambda_N,N)}(\varphi) \tag{3.1}$$

$$V_0^{(\Lambda_N,N)}(\varphi) = \frac{1}{2}\left(\tilde{Z}(N)-1\right)\int d^2x\, |\nabla\varphi|^2 + \Lambda_N^2 \frac{\lambda_0}{2} g_0^2(N) \int d^2x \left(\tilde{Z}\,|\varphi(x)|^2 - \frac{1}{g_0^2(N)}\right)^2 \tag{3.2}$$

Here $\Lambda_N = L^N$ (in mass units) is the U.V. cut off, i.e. $\Lambda^N \underset{N\to\infty}{\longrightarrow} \infty$, $L > 1$. We have introduced a Pauli-Villars cut off. λ_0 is a free dimensionless parameter (the dimensionless σ-mass). In Section 2 we have seen that by adding a magnetic field (to avoid perturbation theory I.R. divergences) and choosing $\tilde{Z}(N)$, $g_0^2(N)$ in the standard way [2] , we recover 1-loop non-linear σ-model results. The fact that the cutoff removal is achieved by letting the dimensional mass^2 $m_\sigma^2 = \Lambda_N^2\, \lambda_0$ with λ_0 arbitrary immediately suggests (by virtue of our knowledge of critical phenomena due to Wilson [4,8]) that under the RG the dimensionless parameter λ is driven upto $O(g^2)$ to a fixed point ("infra-red stable"). In this sction we will verify this point as well as how asymptotic freedom is obtained by making some approximations to the full Wilson RG.

Remark

We shall work with the symmetric form of the theory i.e. start from (3.1), (3.2). To avoid I.R. divergences we should put a space-time volume cutoff. However as we shall see (and as is well-known) in the Wilson RG transformation no I.R. divergences are met. Hence the volume cutoff is ignored, although it is not difficult to keep track of it (F.J. Wegner [9]). All parameters in (3.2) are dimensionless. Hence by rescaling

$$\langle \varphi(x_1)\cdots\varphi(x_{2K})\rangle_{S_0^{(\Lambda_N,N)}} = \langle \Phi(L^N x_1)\cdots\Phi(L^N x_{2K})\rangle_{S_0^{(N)}} \tag{3.3}$$

where

$$S_0^{(N)} = \int d^2x \left| e^{-\frac{\Delta}{2}} \nabla\Phi \right|^2 + V_0^{(N)}(\Phi)$$

$$V_0^{(N)}(\Phi) = \frac{1}{2}\left(\tilde{Z}(N)-1\right)\int d^2x\, |\nabla\Phi|^2 + g_0^2(N)\frac{\lambda_0}{2}\int d^2x \left(\tilde{Z}\,|\Phi(x)|^2 - \frac{1}{g_0^2(N)}\right)^2 \tag{3.4}$$

The free part now has a "unit cutoff" and $\tilde{Z}$, g_0 carry the cutoff dependence.

Hold the points of (3.3) non-coincident : $|x_i - x_j| \gg L^{-M}$ (M fixed). L^M will be our reference momentum scale. In the Wilson method [4] at each RG step we lower the cutoff from 1 to $L^{-1}1$ and rescale back to 1. After (N-M) steps. We get :

$$\langle \varphi(x_1)\cdots\varphi(x_{2K})\rangle_{S_0^{(\Lambda_N,N)}} = \langle \Phi(L^M x_1)\cdots\cdots \Phi(L^M x_{2K})\rangle_{S^{(N)}_{N-M}} \tag{3.5}$$

$$S^{(N)}_{N-M} = \int d^2x \, \left|e^{-\frac{\Delta}{2}} \nabla \Phi\right|^2 + V^{(N)}_{N-M}(\Phi)$$

$$V^{(N)}_{\ell} \underset{RG}{\rightarrow} V^{(N)}_{\ell+1} \tag{3.6}$$

In (3.5) we have ignored a slowly varying finite (N-independent) momentum dependent factor. The idea is now to choose Z(N), $g_0(N)$ such that

$$\lim_{\substack{N\to\infty \\ M \text{ fixed}}} V^{(N)}_{N-M} = \tilde{V}_M \quad \text{exists} \tag{3.7}$$

in which case $\tilde{V}_M$ is the <u>renormalized trajectory</u> (L^M is the renormalization scale). For our purposes here we adopt a continuous form of the Wilson RG (Wilson-Kogut [4], Wegner [9]). We lower the cutoff continuously from 1 to $e^{-t}1$ and scale back to 1. Starting from t = 0 we integrate out to t_{N-M}=(N-M) log L and the renormalized trajectory is

$$\lim_{\substack{N\to\infty \\ M \text{ fixed}}} V^{(N)}_{t_{N-M}} = \tilde{V}_M \tag{3.8}$$

Under the continuous Wilson RG, $V_0^{(N)} \rightarrow V_t^{(N)}$ given by :

$$e^{-V_t^{(N)}(\Phi)} = \int d\mu_{C_t^h}(\zeta)\, e^{-V_0^{(N)}(\Phi(e^{-t}\cdot)+\zeta)} \tag{3.9}$$

where $\mu_{C_t^h}$ is a Gaussian measure with covariance :

$$C_t^h(x-y) = \int \frac{d^2k}{(2\pi)^2}\, e^{ik\cdot(x-y)} \frac{\left(e^{-k^2} - e^{-k^2/e^{-2t}}\right)}{k^2} \tag{3.10}$$

Note that the hard propagator C_t^h has <u>no</u> I.R. divergences, so that in (3.9) no I.R. divergences occur. From (3.9-3.10) we easily drive the non-linear functional differential equation, (Wilson [4], Wegner [9], see also Polchinski [10])

$$\frac{\partial V_t^{(N)}}{\partial t} = -\sum_{i=1}^{n} \int d^2x \left(x \cdot \nabla \Phi_i(x)\right) \frac{\delta V_t^{(N)}}{\delta \Phi_i(x)}$$
$$- \int d^2x\, d^2y\, K(x-y) \sum_{i=1}^{n} \left\{ - \frac{\delta^2 V_t^{(N)}}{\delta \Phi_i(x) \delta \Phi_i(y)} + \frac{\delta V_t^{(N)}}{\delta \Phi_i(x)} \frac{\delta V_t^{(N)}}{\delta \Phi_i(y)} \right\} \quad (3.11)$$

$$K(x-y) = \int \frac{d^2q}{(2\pi)^2}\, e^{iq(x-y)}\, e^{-q^2} \quad (3.12)$$

Because of the non-local kernel K(x-y), the third term in (3.11) generates non-local contributions to the effective potential $V_t^{(N)}$. This is also the source of the wave function renormalization. As a first step in the full analysis we make the local approximation (for a recent study, see Felder [11]) to the RG equation (3.11). This consists in replacing K(x-y) where it sits in the third term of (3.11) by $\delta^{(2)}$ (x-y). This has the disavantage that we miss out the wave function renormalization which will be recovered by going beyond the local approximation in the second step of the analysis. However it has the great advantage of giving a purely local evolution under the RG which can be throughly analyzed and gives a very interesting picture of what is essentiallly going on.

In the above local approximation the evolved potential V_t remains local (we drop the superscript "N" which refers to the initial condition):

$$V_t(\Phi) = \int d^2x\; v_t\left(|\Phi(x)|^2\right) \quad (3.13)$$

and calling $\quad |\Phi(x)|^2 = \tau$

$$\frac{\partial v_t}{\partial t} = 2v_t + \frac{n}{2\pi} v_t' + 4\tau\left(\frac{1}{2(2\pi)} v_t'' - (v_t')^2\right) \quad (3.14)$$

where v'_t, v''_t are first and second derivatives with respect to τ. This is a 2-dimensional non-linear partial differential equation. In the local approximation, no wave functional renormalization is generated and hence we can set $\tilde{Z} = 1$, so that (from (3.4)) our initial local potential is:

$$v_0(\tau) = g_0^2 \frac{\lambda_0}{Z}\left(\tau - \frac{1}{g_0^2}\right)^2 \quad (3.15)$$

Here $\frac{1}{g_0^2}$ is the minimum of $v_0(\tau)$ (its dependence on the UV cutoff N wil be fixed later) and $\lambda_0 g_0^2$ is the curvature at the minimum which defines λ_0. In the following λ_0 is an N-independent free positive parameter. We now define $\frac{1}{g_t^2}$ as the minimum of the evolved potential $v_t(\tau)$. Define

$$\hat{v}_t(\tau) = v_t\left(\tau + \frac{1}{g_t^2}\right) \tag{3.16}$$

Then $\hat{v}_t$ has minimum at $\tau = 0$. The differential equation (3.14) gives

$$\frac{\partial \hat{v}_t}{\partial t} = 2\hat{v}_t + \frac{n}{2\pi}\hat{v}_t' + 4\left(\tau + \frac{1}{g_t^2}\right)\left(\frac{1}{2}\cdot\frac{1}{2\pi}\hat{v}_t'' - (\hat{v}_t')^2\right) - \left(\frac{dg_t^2}{dt}\right)\frac{1}{(g_t^2)^2}\hat{v}_t' \tag{3.17}$$

Taking another derivative of (3.17) w.r. to τ, and demanding that $\hat{v}'_t(0) = 0$ gives us :

$$\frac{dg_t^2}{dt} = \frac{1}{2\pi}\left[(n+2) + \frac{2}{g_t^2}\frac{\hat{v}_t'''(0)}{\hat{v}_t''(0)}\right](g_t^2)^2 \tag{3.18}$$

Let us expand out $\hat{v}_t$ as a power series in τ, upto $O(\tau^6)$.

$$\hat{v}_t(\tau) = g_t^2\frac{\lambda_t}{2}\tau^2 + (g_t^2)^2\frac{w_t}{3!}\tau^3 + (g_t^2)^3\eta_t\tau^4 + (g_t^2)^4\frac{\nu_t}{5!}\tau^5 + (g_t^2)^5\frac{\rho_t}{6!}\tau^6 + O\left((g_t^2)^{5+\delta}\tau^{6+\delta}\right) \tag{3.19}$$

The linear term is always absent because of (3.18). We need to go upto the τ^5 term to get the asymptotic freedom formula upto order $(g^2)^3$ (i.e. 2 loops) correctly. We have nevertheless kept the τ^6 term to show that it and (also beyond) does not affect the renormalization analysis.(3.18) can now be written :

$$\frac{dg_t^2}{dt} = \frac{1}{2\pi}\left[(n+2) + \frac{2w_t}{\lambda_t}\right](g_t^2)^2 \qquad (3.20)$$

From (3.17), (3.19) and using (3.20) we get the following flow equations for the parameters in (3.19):

$$\frac{d\lambda_t}{dt} = 2\lambda_t(1-4\lambda_t) - \frac{g_t^2}{2\pi}\left\{\left[(n+2)+\frac{2w_t}{\lambda_t}\right] - w_t\left(1-\frac{w_t}{\lambda_t}\right) - 2\eta_t\right\} \qquad (3.21)$$

$$\frac{dw_t}{dt} = 2\,(1-12\lambda_t)w_t - (4!)\lambda_t^2 - \frac{2g_t^2}{2\pi}\left\{\left[(n+2)+\frac{2w_t}{\lambda_t}\right] - \eta_t\left(2-\frac{w_t}{\lambda_t}\right) - \nu_t\right\} \qquad (3.22)$$

$$\frac{d\eta_t}{dt} = 2\eta_t(1-16\lambda_t) - 24w_t^2 + O(g_t^2) \qquad (3.23)$$

$$\frac{d\nu_t}{dt} = 2\nu_t(1-20\lambda_t) - 30w_t^2 - \frac{5!\times 4}{3}\lambda_t\eta_t - 80w_t\eta_t + O(g_t^2) \qquad (3.24)$$

Finally,

$$\frac{d\rho_t}{dt} = 2\rho_t(1-30\lambda_t) - 120\left(2\lambda_t\nu_t + 4w_t\eta_t + w_t\nu_t\right) - 80\eta_t^2 + O(g_t^2) \qquad (3.25)$$

To obtain the flow (3.20) for $(g^2{}_t)$ exactly to order $(g^2{}_t)^3$, (2-loop information), we need W_t, λ_t exactly to order $g^2{}_t$. From (3.21-3.22) we see that we need η_t, ν_t to O(1), and ρ_t (the τ^6 term) plays no role for this part of the calculation.

Let $\hat{\lambda}_t$, $\hat{W}_t$, $\hat{\eta}_t$, $\hat{\nu}_t$, $\hat{\rho}_t$ denote the solutions of the approximate flow equations obtained from (3.21-3.25) by putting $g^2 = 0$. As $t \to \infty$, these parameters tend to fixed point values ($\hat{\lambda}_*$, $\hat{W}_*$, $\hat{\eta}_*$, $\hat{\nu}_*$, $\hat{\rho}_*$).

In particular $\hat{\lambda}_* = \frac{1}{4}$, $\hat{w}_* = -\frac{3}{8}$, $\hat{\eta}_* = -\frac{9}{16}$, $\hat{\nu}_* = \frac{45}{256}$

This for large t we can replace (3.21-3.22) by,

$$\frac{d\lambda_t}{dt} = 2\lambda_t\,(1-4\lambda_t) + a_*\, g_t^2 + O(g_t^4)$$

$$\frac{dw_t}{dt} = 2\,(1-12\lambda_t)\,w_t - (4!)\,\lambda_t^2 + b_*\, g_t^2 + O(g_t^4) \qquad (3.26)$$

where a_*, b_* contain the above fixed point information. From (3.26) it is easy to see tha for large t

$$\lambda_t \to \hat{\lambda}_* + c_*\, g_t^2 + O(g_t^4)$$

$$w_t \to \hat{w}_* + d_*\, g_t^2 + O(g_t^4) \qquad (3.27)$$

Plugging in (3.27) in (3.20) for large t we obtain :

$$\frac{dg_t^2}{dt} = \frac{(n-1)}{2\pi}(g_t^2)^2 + e_*\,(g_t^2)^3 + O((g_t^2)^4) \qquad (3.28)$$

<u>The discrepancy in the leading term from its true value</u> $(n-2)/(2\pi)$ may be attributed to the local approximation.

The initial value : $g_0^2(N)$ is now fixed by taking the two loop formula

$$\frac{dg_t^2}{dt} = \frac{(n-1)}{2\pi}(g_t^2)^2 + e_*\,(g_t^2)^3 \qquad (3.29)$$

and demanding that for $t_{N-M} = (N-M)\log L$, $g^2{}_{N-M}$ is held fixed as $\tilde{g}^2{}_M$ (normalization condition at L^M, M fixed). With this choice of $g_0^2(N)$ one obtains that the solution $g^2{}_{N-M}$ of (3.28) (with three loop and higher terms) stabilize as $N \to \infty$, to $\bar{g}^2{}_M < \infty$. This phenomenon is well known (P.K. Mitter learnt this in 1978 from the late K. Symanzik in the context of Creutz' simulations). The initial $g_0^2(N)$ being thus fixed, we go back to (3.26) and see that for large t ($\sim (N-M)\log L$), (3.26) settles down to

$$\frac{d\lambda_t}{dt} = 2\lambda_t\,(1-4\lambda_t) + a_*\,\bar{g}_M^2 + O(\bar{g}_M^4)$$

$$\frac{dw_t}{dt} = 2\,(1-12\lambda_t)\,w_t - (4!)\,\lambda_t^2 + b_*\,\bar{g}_M^2 + O(\bar{g}_M^4) \qquad (3.30)$$

Letting now $t = t_{N-M} = (N-M) \log L \underset{N\to\infty}{\to} \infty$, we see that λ_t, w_t approach "fixed points"

$$\lambda_*(M) = \hat{\lambda}_* + O(\bar{g}_M^2)$$
$$w_*(M) = \hat{w}_* + O(\bar{g}_M^2) \tag{3.31}$$

A similar discussion holds for (3.23-3.25)

As a consequence we obtain as $N \to \infty$, the renormalized trajectory :

$$\tilde{v}_M(|\Phi|^2) = \tilde{\hat{v}}_M\left(|\Phi|^2 - \frac{1}{\bar{g}_M^2}\right) \tag{3.32}$$

where

$$\tilde{\hat{v}}_M(\tau) = \lim_{N\to\infty} \hat{v}_{t_{N-M}}(\tau)$$

$$= \bar{g}_M^2 \frac{\lambda_*(M)}{2}\tau^2 + (\bar{g}_M^2)^2 \frac{w_*(M)}{3!}\tau^3 + (\bar{g}_M^2)^3 \frac{\eta_*(M)}{4!}\tau^4$$
$$+ (\bar{g}_M^2)^4 \frac{\nu_*(M)}{5!}\tau^5 + O((\bar{g}_M^2)^5 \tau^6) \tag{3.33}$$

where $\lambda_*(M) = \hat{\lambda}_* + O(\bar{g}_M^2)$ etc.

We note that g is the only marginal variable, and otherwise all other parameters contract to fixed point values under RG.

The above discussion is clearly deficient since it is based on the "local approximation" to the RG : thus the wave function renormalization is missed out, and the numerical coefficients in the asymptotic freedom formula are inaccurate. Moreover (3.33) is just a weak coupling expansion (in $\bar{g}_M$) presumably valid for small fields. Nevertheless it explains (we hope clearly) the mechanisms at work. A more detailed study will be published elsewhere [12]. The above picture that we have derived is similar to (but not all details are the same as) that obtained by Gawedski and Kupiainen [13] in a study of a hierarchichal model.

4° THE WILSON APPROXIMATE RECURSION FORMULA

This recursion is $\omega \to \omega'$ where

$$e^{-\omega'(x)L^{-2}} = \text{cnst} \int e^{-\left(\frac{\omega(x+y)+\omega(x-y)}{2}\right) - y^2/2}\, dy \tag{4.1}$$

and L > 1. This can be regarded either as an approximation to the full renormalization group transformation as in [4] or as the renormalization group transformation for a certain hierarchical model. Note that it differs from the recursion used by Gawedzki-Kupainen [13]. While comparing with our earlier discussion the identification that is made is

(i) We choose $\tilde{Z}=1$ (in the hieararchical approximation there is no true wave-function renormalization)

(ii) $\frac{m_\sigma^2 g_0^2 a^2}{8}\left(|\phi|^2-\frac{1}{g_0^2}\right)^2 \rightarrow \omega_0(|\phi|^2)$, where by ω_0 we denote the initial point for the recursion.

We shall consider O(n) invariant ω. This ondition is clearly preserved by the recursion. We write

$$\omega(\phi) = n\,\hat{v}\left(|\phi|^2-\frac{1}{g^2}\right) \tag{4.2}$$

where $\hat{v} \geqslant 0$, v(0) = 0 (this defines g if $\hat{v}$ has a unique minimum which is taken at the origin -this will be the case in our approximate considerations. In general a more rigorous definition of g will have to be given). Then the recursion (4.1) become $\hat{v} \rightarrow \hat{v}'$

$$e^{-nL^{-2}\hat{v}'(\tau+\varepsilon)} = \mathrm{cnst}\int_{-\infty}^{\infty} d\zeta \int_0^{\infty} e^{-\frac{ns}{2}}\, s^{\frac{n-3}{2}}\, e^{-\frac{n}{2}g^2\zeta^2}\cdot$$
$$\cdot \exp -\frac{n}{2}\Big[\hat{v}\left(\tau+s+|\zeta|^2g^2+2\sqrt{g^2\tau+1}\,\zeta\right) \tag{4.3}$$
$$+\,\hat{v}\left(\tau+s+\zeta^2g^2-2\sqrt{g^2\tau+1}\,\zeta\right)\Big]\,ds$$

where $\varepsilon = \frac{1}{g^2}-\frac{1}{(g')^2}$, and $\frac{1}{(g')^2}$ is the minimum of ω'

We can put g = 0 in the above recursion keeping ε fixed, to obtain

$$e^{-n\hat{v}'(\tau+\varepsilon)L^{-2}} = \mathrm{cnst}\int_{-\infty}^{\infty} d\zeta \int_0^{\infty} ds\, e^{-\frac{ns}{2}}\, s^{\frac{n-3}{2}}\cdot$$
$$\cdot \exp -\frac{n}{2}\left[\hat{v}(\tau+s+2\zeta)+\hat{v}(\tau+s-2\zeta)\right] \tag{4.4}$$

We have proved (see the Appendix) that for n=3, there exists an $\varepsilon=\varepsilon^* \geqslant 0$ such that this recursion possesses a fixed point $\hat{v}_3^*$. This proof, however, gives very little information. A saddle point computation shows that in the n = ∞ limit the above transformation has a fixed point $\hat{v}_\infty^*$, with $\varepsilon^* = 1$. This is done closely following Ma [14] and yields $\frac{d\hat{v}_\infty^*}{d\tau} = f^{-1}$ where f is the function

$$f(u) = \sum_{m=0}^{\infty} \frac{2u}{L^{2(m+1)}+2u} \tag{4.5}$$

A full analysis of the scaling fields is possible and shows that this fixed point is attractive.

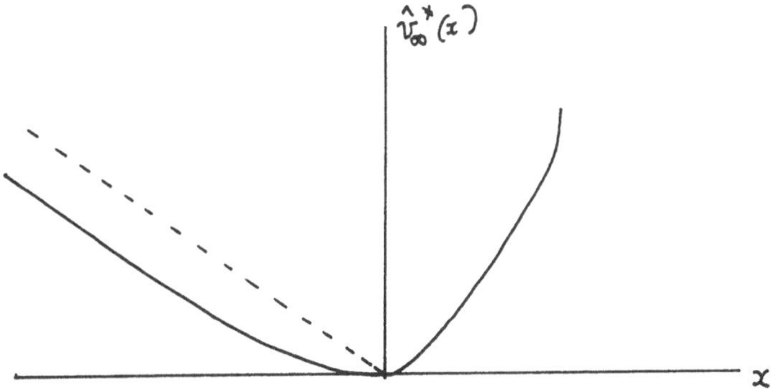

Figure 4 - A sketch of $\hat{v}_\infty^*$

We have also studied (4.4) (for n=3) numerically. This confirms the n = ∞ picture.

We have studied the recursions (4.3) and (4.4) to order 1/n. We get

$$\varepsilon^* = 1 + \frac{1}{n}\left\{-3 + \frac{\left(2 + \frac{\hat{v}_\infty^{*(3)}(0)}{\hat{v}_\infty^{*(2)}(0)}\right)}{1+\hat{v}_\infty^{*(2)}(0)} + \frac{1}{2}\,\frac{\hat{v}_\infty^{*(3)}(0)}{\hat{v}_\infty^{*(2)}(0)}\right\} \tag{4.6}$$
$$+\, O\left(\frac{1}{n^2}\right)$$

where $\hat{v}_\infty^{*(i)}$ denotes the ith derivative of $\hat{v}_\infty^*$. Continuing this way one gets what we believe is an asymptotic series for ε^*. Note, howeve that we do <u>not</u> expect an exact formula for ε^*

One might attempt to take a scaling limit of the hierarchical theory by letting

$$\frac{m_\sigma^2 g_0^2 a^2}{g} \xrightarrow[a \to 0]{} \tilde{\lambda}^* \tag{4.7}$$

$$\frac{1}{g_0^2(a)} = \frac{1}{g^2} - \varepsilon^* \log a + O(g^2) \tag{4.8}$$

where $\tilde{\lambda}^*$ is the coefficient of the expansion of $\hat{v}_n^*$ around 0, i.e., $\hat{v}_n^*(\tau) = \tilde{\lambda}^* \tau^2 + O(\tau^3)$ In fact this programme can be carried out to order 1/n. However, note that the coefficients of the asymptotic freedom formula refer to the fixed point $\hat{v}_n^*$ and are hence not exactly known.

Finally we remark that in the variable λ (cf., [1.3]) we do not find a fixed point. This is in contrast to sections 2 and 3 and points to the fact that in two dimensions hierarchical models cannot be wholly trusted.

REFERENCES

[1] A.M. Polyakov, Phys. Lett. 59 B, 79 (1975).

[2] E. Brézin and J. Zinn-Justin, Phys. Rev. B 14, 3110 (1976).

[3] E. Brézin, J.C. Le Guillou and J. Zinn-Justin, Phys. Rev. D 14, 2615 (1976).

[4] K.G. Wilson and J. Kogut, Phys. Rep. 12 C, 75 (1974).

[5] J. Glimm and A. Jaffe, Fortschritte der Physik 21, 327 (1973).

[6] K. Gawedzki and A. Kupiainen, Comm. Math. Phys. 102, 1 (1985).

[7] J. Feldman, J. Magnen, V. Rivasseau, R. Sénéor, Comm. Math. Phys. 103, 67(1986).

[8] K.G. Wilson, Phys. Rev. D7, 2911 (1973).

[9] F.J. Wegner, in "Phase Transitions and Critical Phenomena, vol. 6, Eds. C. Domb and M.S. Green, Acad. Press, London, New-York 1976.

[10] J. Polchinski, Nucl. Phys. B231, 269 (1984).

[11] G. Felder, IHES, Bures-sur-Yvette, preprint 1986.

[12] P.K. Mitter and T.R. Ramadas, in preparation.

[13] K. Gawedzki and A. Kupiainen, Comm. Math. Phys. 106, 533 (1986).

[14] S. Ma, "Modern Theory of Critical Phenomena, New-York, Benjamin 1976.

[15] P.K. Mitter and P.H. Weisz, Phys. Rev. D8, 4410 (1973).

[16] D. Gross and A. Neveu, Phys. Rev. D10, 3235 (1974).

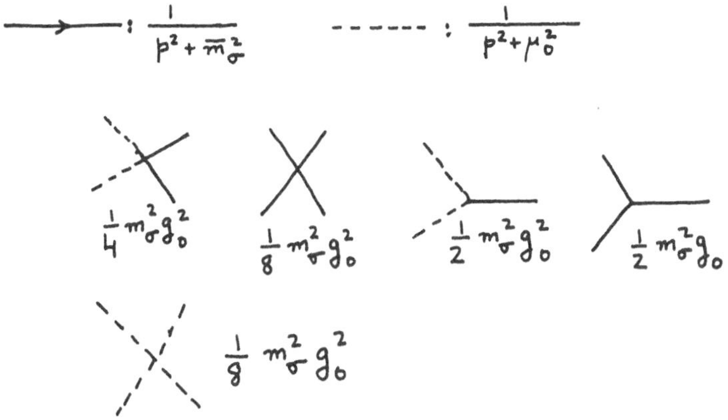

Figure 1 - The basic propagator and vortices

$$G_\Lambda^{(2)}(x,y) \equiv \langle \pi(x) \cdot \pi(y) \rangle$$

$$= (n-1) \; \text{-----}$$

$$- \frac{g_0^2 m_\sigma^2}{2} \left\{ [2(n-1)+(n-1)^2] \; \text{---} \!\!\!\!\!\!\!\!\!\!\!\! \text{---} \quad + (n-1) \; \text{---} \!\!\!\! \text{---} \right\}$$

$$+ \frac{g_0^2 m_\sigma^4}{2} \left\{ 2(n-1) \; \text{---} \!\!\!\! \text{---} \quad + (n-1)^2 \; \text{---} \!\!\!\! \text{---} \quad + 3(n-1) \; \text{---} \!\!\!\! \text{---} \right\}$$

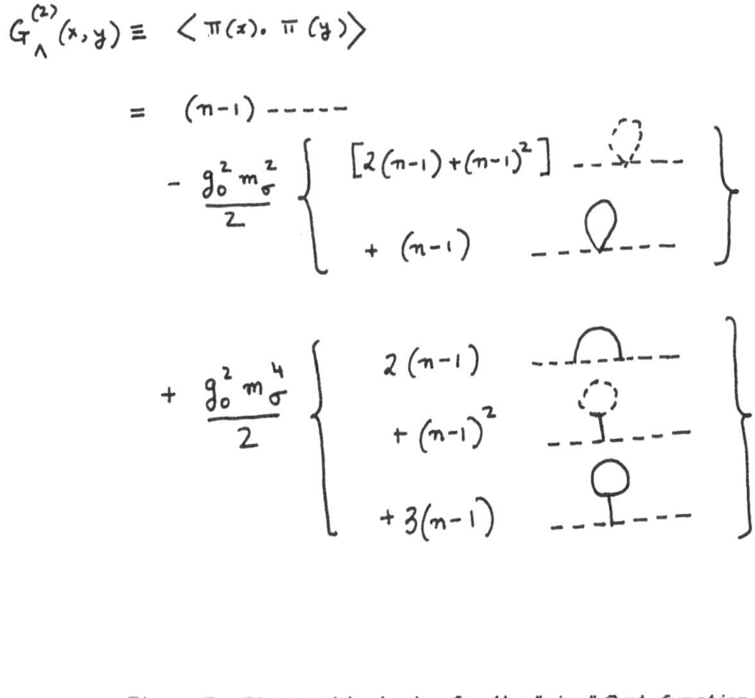

Figure 2 - The combinatories for the "pion" 2 pt. function

Figure 3 - The 1-π irreducible proper 4 point function. (external legs deleted everywhere, but so indicated only in the firs diagram.

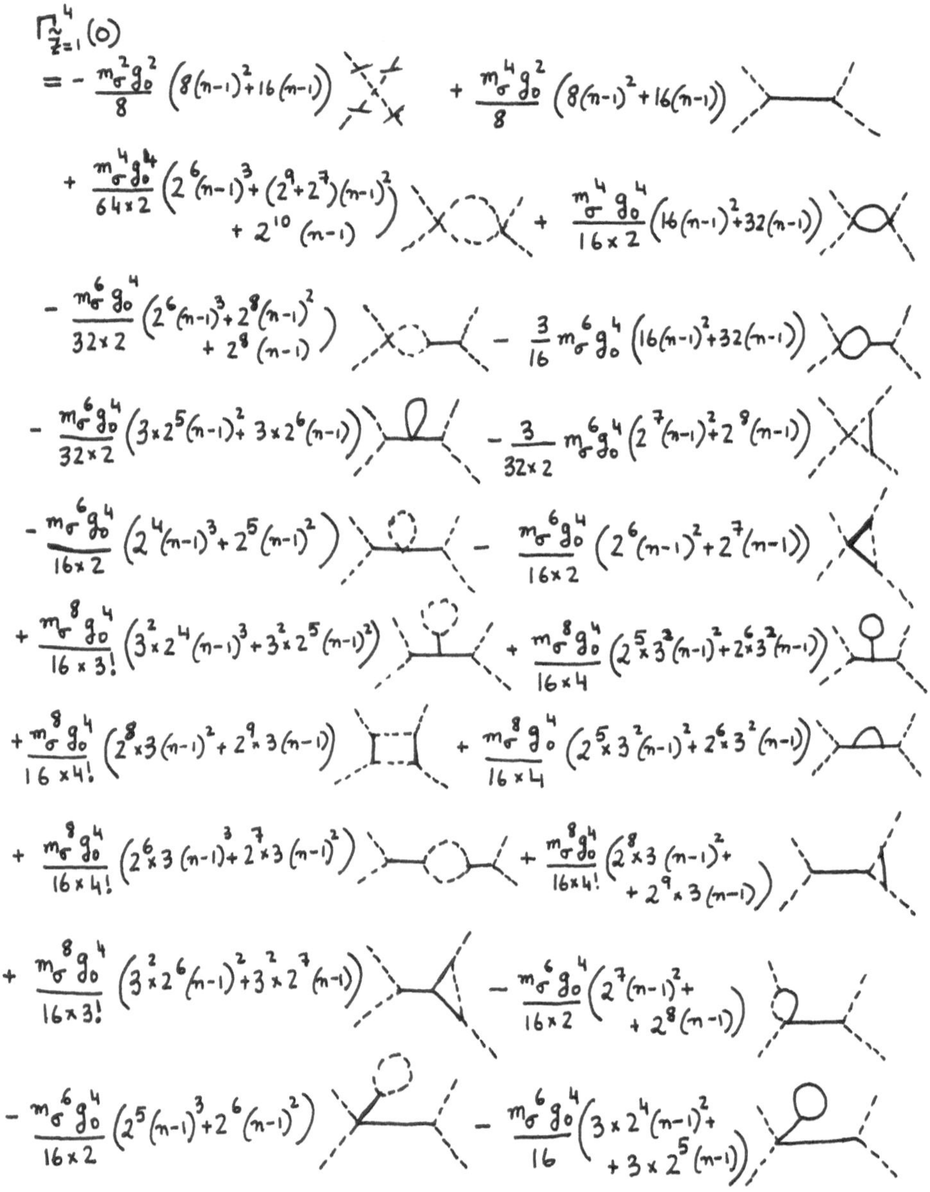

$$\Gamma^4_{\tilde{z}=1}(0) = -\frac{m_\sigma^2 g_0^2}{8}\left(8(n-1)^2+16(n-1)\right) + \frac{m_\sigma^4 g_0^2}{8}\left(8(n-1)^2+16(n-1)\right)$$

$$+\frac{m_\sigma^4 g_0^4}{64\times 2}\left(2^6(n-1)^3+(2^9+2^7)(n-1)^2 + 2^{10}(n-1)\right) + \frac{m_\sigma^4 g_0^4}{16\times 2}\left(16(n-1)^2+32(n-1)\right)$$

$$-\frac{m_\sigma^6 g_0^4}{32\times 2}\left(2^6(n-1)^3+2^8(n-1)^2 + 2^8(n-1)\right) - \frac{3}{16} m_\sigma^6 g_0^4\left(16(n-1)^2+32(n-1)\right)$$

$$-\frac{m_\sigma^6 g_0^4}{32\times 2}\left(3\times 2^5(n-1)^2+3\times 2^6(n-1)\right) - \frac{3}{32\times 2} m_\sigma^6 g_0^4\left(2^7(n-1)^2+2^9(n-1)\right)$$

$$-\frac{m_\sigma^6 g_0^4}{16\times 2}\left(2^4(n-1)^3+2^5(n-1)^2\right) - \frac{m_\sigma^6 g_0^4}{16\times 2}\left(2^6(n-1)^2+2^7(n-1)\right)$$

$$+\frac{m_\sigma^8 g_0^4}{16\times 3!}\left(3^2 2^4(n-1)^3+3^2 2^5(n-1)^2\right) + \frac{m_\sigma^8 g_0^4}{16\times 4}\left(2^5\times 3^2(n-1)^2+2^6\times 3^2(n-1)\right)$$

$$+\frac{m_\sigma^8 g_0^4}{16\times 4!}\left(2^8\times 3(n-1)^2+2^9\times 3(n-1)\right) + \frac{m_\sigma^8 g_0^4}{16\times 4}\left(2^5\times 3^2(n-1)^2+2^6\times 3^2(n-1)\right)$$

$$+\frac{m_\sigma^8 g_0^4}{16\times 4!}\left(2^6\times 3(n-1)^3+2^7\times 3(n-1)^2\right) + \frac{m_\sigma^8 g_0^4}{16\times 4!}\left(2^8\times 3(n-1)^2 + 2^9\times 3(n-1)\right)$$

$$+\frac{m_\sigma^8 g_0^4}{16\times 3!}\left(3^2\times 2^6(n-1)^2+3^2\times 2^7(n-1)\right) - \frac{m_\sigma^6 g_0^4}{16\times 2}\left(2^7(n-1)^2 + 2^8(n-1)\right)$$

$$-\frac{m_\sigma^6 g_0^4}{16\times 2}\left(2^5(n-1)^3+2^6(n-1)^2\right) - \frac{m_\sigma^6 g_0^4}{16}\left(3\times 2^4(n-1)^2 + 3\times 2^5(n-1)\right)$$

APPENDIX

The recursion (4.4) for n = 3, L = 2

§1. We denote $g(\tau) = e^{-\frac{3\hat{v}(\tau)}{4}}$ Then (4.4) becomes

$$g'(\tau+\varepsilon) = \text{const} \int_{-\infty}^{\infty} dt \int_0^{\infty} ds \; e^{-\frac{3s}{2}} g^2(\tau+s+t)\, g^2(\tau+s-t) \qquad \text{(A-1)}$$

If we define $s = e^{\frac{3x}{2}}$, $f(s) = g(x)e^{-\frac{x}{2}}$ we get [c = "varying constant"]

$$\frac{d}{ds} f'(\alpha s) = -c \int_0^{\infty} \frac{f^2(su)\, f^2(su^{-1})}{u}\, du \;, \quad f'(+\infty) = 0 \qquad \text{(A-2)}$$

Where $\alpha = e^{3\varepsilon/2}$. We shall now change the definition of c and α in an inessential way. Note that f' is a decreasing function, $f' \to 0$ as $x \to \infty$, and $f'(0) = \frac{c}{2}\left(\int_0^{\infty} \frac{f^2(u)}{u^{1/2}} du\right)$. We define c s.t. $f'(0) = 1$ and α s.t. $f'(1) = a$

for some $0 < a < 1$ fixed once and for all.

Let $K \gg 1$, and define

$\mathcal{F} = \{ f \in C[0,K] \,, f(0)=1 \,, f(1) = a \,, f$ non increasing$\}$

This is a closed convex set in c[0,1]

§2. We define a transformation $f \to Tf = \tilde{f}$ on $\mathcal{F}$. This is done by regarding f as a function on $[0,\infty)$, performing the transformation $f \to f'$ as in §1, and defining $\tilde{f} = \chi_K f'$ where χ_K is the characteristic function of [0,K].

§3. The Leray-Schauder Fixed Point Theorem applies to the transformation T, and we obtain a fixed point f_K^*

§4. We get a priori estimates on α_K^* and f_K^*

$$c(a) < \alpha_K^* < 1$$
$$f_K^*(y) \le c y^{-1/3} \qquad \text{(A-3)}$$
$$-\frac{d f_K^*(y)}{dy} \le c y^{-1/3}$$

We can now get a fixed point of (A-2) by taking a weak limit of the f_K^*

§5. We have very little information about the fixed point g^*. We can, however show $\liminf_{x \to \infty} g^*(x) = 0$

NONLINEAR σ-MODELS WITH BOUNDARY AND OPEN STRINGS

H. Dorn, H. J. Otto
Sektion Physik, Humboldt-Universität Berlin, DDR
Invalidenstraße 42, Berlin 1040

Abstract:

We present a short review of some perturbative results concerning the renormalization of nonlinear σ-models with boundary relevant to the description of open bosonic strings. Special emphasis is given to the so-called β-function approach to derive effective actions for the massless excitations of the string. We comment also on the correct definition of the 2 dimensional stress tensor in the presence of antisymmetric tensor field background.

1. Introduction

Relativistic strings describe as its excitations an infinite tower of usual point particles. The corresponding mass scale is the Planck mass $1/\alpha'$. Usual physics appears to be that of the massless excitations. Therefore, one is interested in an effective action for the massless fields derived via an α' expansion from string theory. In this sense string theory can provide a guiding principle to reach consistent anomaly free unified field theories. Typical string induced effects in such theories are e. g. the Chern-Simons completion of the field strength of the antisymmetric tensor field which is crucial for anomaly concellation / 1 /, higher curvature terms for gravity and its supersymmetric extensions / 2 /, and structures of nonlinear electrodynamics and its non-Abelian generalization / 3 /. There are several methods to derive the effective action, and within the inherent ambiguities due to renormalization scheme dependence and/or the freedom to redefine the fields all of them yield the same result / 4, 5 /.

In this talk we restrict ourselves to the β-function approach which is based on the requirement of conformal invariance of the generalized nonlinear σ-model describing the motion of the string in background fields corresponding to its own massless excitations. This procedure we demonstrate for open bosonic strings / 6, 7 /. The limitation to bosonic strings is made for simplicity. Although closed strings appear to have better phenomenological prospects open strings still are not ruled out and in particular due to its simple coupling mechanism to background gauge fields allow to some extent statements in arbitrary order of α' / 7, 8 /. They seem to be also a suitable testing ground for ideas concerning an interference of usual ultraviolet divergencies and divergencies emerging in limiting cases of the topology of the string world surfaces / 9, 10 /.

Our σ-model describes an open string $x^{\mu}(z)$ in the background fields corresponding to the massless excitations of open and closed strings: gauge field $A_{\mu}(x)$, gravitational field $G_{\mu\nu}(x)$, antisymmetric tensor field $B_{\mu\nu}(x)$ and dilaton $\phi(x)$.

$$S = S_M + S_{\partial M}$$

$$S_M = \frac{1}{4\pi\alpha'} \int_M d^2z \left(\sqrt{g}\, g^{mn} G_{\mu\nu}(x(z))\, \partial_m x^{\mu}(z)\, \partial_n x^{\nu}(z) - \right.$$
$$\left. - i\epsilon^{mn} B_{\mu\nu}(x)\, \partial_m x^{\mu} \partial_n x^{\nu} + \alpha' \sqrt{g}\, R^{(2)} \phi(x) \right) \tag{1}$$

$$S_{\partial M} = - \log \operatorname{tr} P \exp i \int_{\partial M} A_{\mu}(x)\, dx^{\mu} - \int_{\partial M} \frac{1}{2\pi} K(s)\, \phi(x(s))\, ds \,.$$

$$(m, n, \ldots = 1, 2 \; ; \; \mu, \nu, \ldots = 1, 2, \ldots, D)$$

g_{mn} is the 2 dimensional metric, $R^{(2)}$ the corresponding curvature scalar and $K(s)$ the external curvature of the boundary ∂M. (We consider g_{mn} positive definite. To reach the pseudo-Riemannian (Minkowski) case by analytic continuation $B_{\mu\nu}$, $G_{\mu\nu}$ and A_{μ} have to be real.) In the Abelian case $S_{\partial M}$ describes the coupling of an external gauge field to opposite charges sitting at the ends of the string. The generalization to Yang-Mills fields naturally leads to the Wilson loop. The term including $K(s)$ ensures the correct coupling of the dilaton $\phi(x)$ to the Euler characteristics / 10 /. From the pure σ-model point of view it is quite natural to couple $x^{\mu}(z)$ to both open (A_{μ}) <u>and</u> closed (G, B, ϕ) string excitations. This coupling is also justified within the string picture / 10 /.

The rest of the talk is organized as follows. In section 2 we scetch the relation between conformal invariance and renormalization group β -functions / 11 - 13 /. Section 3 reviews our own calculation of β -functions in lowest orders / 6 /. Section 4 discusses the progress made in summing the α' series / 7, 8 / and the attempt to include effects of higher topologies of the parameter domain M (string loops) / 9, 10 /. Finally, section 5 gives a preliminary account of some results concerning the definition of the 2 dimensional stress tensor in the presence of $B_{\mu\nu}(x)$.

2. β -functions and conformal invariance

We treat the renormalization of the nonlinear σ -model given by (1) in the generalized sense of Friedan / 14 /, i. e. all counterterms are classified according to their two and D-dimensional structure and e. g.

$$\int_M d^2z \sqrt{g}\, g^{mn} K_{\mu\nu}(x)\, \partial_m x^\mu \partial_n x^\nu \quad , \qquad \int_{\partial M} L_\mu(x)\, dx^\mu$$

is interpreted as a renormalization of the D-dimensional metric $G_{\mu\nu}$ and the gauge field A_μ, respectively. Then one defines (μ renormalization scale)

$$\beta^G_{\kappa\lambda} = \mu \frac{\partial}{\partial\mu} G_{\kappa\lambda} \Big|_{G^0, B^0, \phi^0, A^0 \text{ fix}} \tag{2}$$

with corresponding eqs. for $\beta^B_{\kappa\lambda}$, β^ϕ and β^A_λ. Due to two and D-dimensional covariance the trace of the 2 dimensional stress tensor has the structure

$$4\pi\alpha' \sqrt{g}\, T^m_{\ m} = \sqrt{g}\, g^{mn} \partial_m x^\mu \partial_n x^\nu \bar\beta^G_{\mu\nu} - i\epsilon^{mn} \partial_m x^\mu \partial_n x^\nu \bar\beta^B_{\mu\nu} + \alpha' \sqrt{g}\, R^{(2)} \bar\beta^\phi$$
$$+ \text{gauge field contribution} + \tag{3}$$
$$+ \text{terms vanishing on shell}$$

with coefficients $\bar\beta$ introduced independent of β at this moment. Besides these two kinds of operator β -functions $\beta(x)$, $\bar\beta(x)$ it is sometimes useful to consider $\tilde\beta$ (x) depending on a background configuration $\tilde{x}(z)$ / 15 /.

$$\langle 4\pi\alpha' \sqrt{g}\, T^m_{\ m} \rangle_{\tilde x} = \sqrt{g}\, g^{mn} \partial_m \tilde{x}^\mu \partial_n \tilde{x}^\nu\, \tilde\beta^G_{\mu\nu}(\tilde x) + \tag{4}$$
$$+ \ldots + \text{nonlocal terms}$$

The most detailed discussion of the relation between β, $\bar\beta$, $\tilde\beta$ is given in refs. / 11, 13 /

$$\bar\beta^{G}_{\mu\nu} = \beta^{G}_{\mu\nu} + 2\alpha' D_{(\mu}\partial_{\nu)}\phi + D_{(\mu}W_{\nu)}$$

$$\bar\beta^{B}_{\mu\nu} = \beta^{B}_{\mu\nu} + S_{\mu\nu}{}^{\lambda}(2\alpha'\partial_\lambda\phi + W_\lambda) + \partial_{[\mu}L_{\nu]} \tag{5}$$

$$\bar\beta^{\phi} = \beta^{\phi} + \alpha'(\partial_\mu\phi)^2 + 1/2\, W^\lambda\partial_\lambda\phi$$

with the field strength for $B_{\mu\nu}$ given by

$$S_{\mu\nu\lambda} = 3/2\,\partial_{[\mu}B_{\nu\lambda]} \quad . \tag{6}$$

The terms containing the quantities W and L are due to operator mixing with total derivative terms and vanish up to 2 loop order / 11, 13 /. The remaining terms in $\bar\beta - \beta$ involve the dilaton field and are due to the explicit breaking of the Weyl invariance $g_{mn} \to \varrho(z)\, g_{mn}(z)$ by the dilaton coupling in (1).

Consistency of the string theory requires $T^m{}_m = 0$ as an operator statement. The resulting conditions $\bar\beta^G = \bar\beta^\phi = \bar\beta^B = 0$ can be interpreted as equations of motions for the background fields.

The whole procedure up to now has been performed for closed strings only. We don't know how to include the gauge field contribution into eq. (3). The main obstacle is the independence of the gauge field coupling in (1) of the 2-dimensional metric. Therefore, for the time being we only can assume that there is a boundary contribution to $T^m{}_m$ parametrized by some $\bar\beta^A_\lambda = \beta^A_\lambda + \Delta\beta^A_\lambda$. An obvious source for a contribution to $\Delta\beta^A_\lambda$ is a boundary term due to the dilaton part of the action. To some extent the situation resembles that of the $B_{\mu\nu}$ part in the stress tensor. The classical stress tensor does not depend on $B_{\mu\nu}$, but there is a $B_{\mu\nu}$-part in the trace anomaly (3), (5). The derivation in refs. / 11 - 13 / is based on dimensional regularization. As we shall see in section 5 there are problems concerning the application of dimensional regularization in the presence of an $\epsilon^{mn}B_{\mu\nu}$ term. Hence further study is required to perform a regularization scheme independent analysis.

3. Lowest order calculations

The β -functions for the fields of the excitations of the closed string are / 16, 15, 13 /

$$\beta^{G}_{\mu\nu} = \alpha'(R_{\mu\nu} - S_{\mu\lambda\rho} S_{\nu}{}^{\lambda\rho}) + O(\alpha'^2)$$

$$\beta^{B}_{\mu\nu} = -\alpha' D^{\lambda} S_{\lambda\mu\nu} + O(\alpha'^2) \tag{7}$$

$$\beta^{\phi} = \frac{D}{6} - \frac{\alpha'}{2} D^{\mu} D_{\mu} \phi - \frac{\alpha'}{6} S_{\mu\nu\lambda} S^{\mu\nu\lambda} + O(\alpha'^2) \quad .$$

Calculations to the next order have been done in refs. / 17, 13 / and in the supersymmetric case up to 4 and 5 loops / 18 /.

We illustrate the method of calculation by a short review of ref. / 6 / where the standard treatment of σ -models without boundary has been extended to the boundary case. The main ingredients are the background field method, the use of Riemann normal coordinates and a careful treatment of boundary terms. The calculation has been performed neglecting ϕ for G, B, A background at 1 loops, pure A background at 2 loop and Abelian A at 3 loop level.

Making a background-quantum split $x^{\mu} \longrightarrow x^{\mu} + y^{\mu}$ and introducing the normal coordinates ξ^{α}

$$y^{\alpha} = \xi^{\alpha} - 1/2\, \Gamma_{\mu\lambda}{}^{\alpha} \xi^{\mu} \xi^{\lambda} + \cdots \tag{8}$$

we get

$$\begin{aligned} S_M[x+y] = S_M[x] + \frac{1}{4\pi\alpha'} \int_M d^2z \, \{ & G_{\alpha\beta}(x)\, \hat{D}_m \xi^{\alpha} \hat{D}_m \xi^{\beta} + \\ & + [(R_{\alpha\mu\beta\nu} + S_{\alpha\mu\lambda} S_{\beta\nu}{}^{\lambda}) \delta^{mn} - i \epsilon^{mn} D_{\alpha} S_{\beta\mu\nu}] \xi^{\alpha} \xi^{\beta} \partial_m x^{\mu} \partial_n x^{\nu} \\ & - i \partial_m [\xi^{\gamma} D_{\gamma} B_{\alpha\beta} \xi^{\alpha} \epsilon^{mn} \partial_n x^{\beta} + B_{\alpha\beta} \xi^{\alpha} \epsilon^{mn} D_n \xi^{\beta}] \} + O(\xi^3) \\ & + \text{term linear in } \xi \, . \end{aligned} \tag{9}$$

D_{α} , $\hat{D}_{\alpha}$ are the covariant derivatives corresponding to the connections $\Gamma_{\alpha\beta\gamma}$, $\Gamma_{\alpha\beta\gamma} + i S_{\alpha\beta\gamma}$ respectively. D_m , $\hat{D}_m$ are their 2 dim. projections. $R_{\alpha\beta\gamma}{}^{\delta} = \partial_{\alpha} \Gamma_{\beta\gamma}{}^{\delta} - \partial_{\beta} \Gamma_{\alpha\gamma}{}^{\delta} + \cdots$, $R_{\beta\gamma} = R_{\alpha\beta\gamma}{}^{\alpha}$.

The expansion of the Wilson loop is based on

$$\begin{aligned} tr\, U[x+y] = tr\, P\Big(U[x] \exp i \int ds \big[& F_{\mu\lambda}\dot{x}^{\lambda} y^{\mu} + 1/2\, D_{\mu_2} F_{\mu_1\lambda}\dot{x}^{\lambda} y^{\mu_1} y^{\mu_2} + \\ & + 1/2\, F_{\mu_1\mu_2} y^{\mu_1}\dot{y}^{\mu_2} + \sum_{n=3}^{\infty}\Big(\tfrac{1}{n!} D_{\mu_1}\cdots D_{\mu_{n-1}} F_{\mu_n\lambda}\dot{x}^{\lambda} y^{\mu_1}\cdots y^{\mu_n} + \\ & + \tfrac{n-1}{n!} D_{\mu_1}\cdots D_{\mu_{n-2}} F_{\mu_{n-1}\mu_n} y^{\mu_1}\cdots \dot{y}^{\mu_n}\Big)\big]\Big) . \end{aligned} \tag{10}$$

As propagator for the quantum field $\xi^A = V^A_\alpha \xi^\alpha$ where V^A_α is the D-dim. vielbein we choose

$$\overbrace{\xi^A(z)\,\xi^B}(z') = 2\pi\alpha'\,\delta^{AB} N(z,z') \tag{11}$$

with N denoting the Neumann function for M. Then we get for the effective action at 1 loop order

$$\begin{aligned} -\Gamma[x] = {} & 1/2 \int_M d^2z\, N(z,z)\,\big[(R_{\alpha\beta} - S_\alpha{}^{\mu\nu} S_{\beta\mu\nu})\,\delta^{mn} + i\, D^{\mu} S_{\mu\alpha\beta}\,\epsilon^{mn}\big]\,\partial_m x^\alpha \partial_n x^\beta \\ & + \frac{i}{2}\int_{\partial M} ds\, N(z(s),z(s))\,\big[D^\alpha B_{\alpha\beta}\dot{x}^\beta - i\,\dot{z}_m \epsilon^{mk}\partial_k x^\beta S^{\alpha\delta}{}_\beta\, B_{\delta\alpha}\big] \\ & + \pi i\alpha' \int_{\partial M} ds\, N(z(s),z(s))\; tr\, P\big(U\big[D^{(A,\Gamma)\alpha} F_{\alpha\beta}\dot{x}^\beta - i\,\dot{z}_m\epsilon^{mk}\partial_k x^\beta S^{\alpha\delta}{}_\beta\, F_{\delta\alpha}\big]\big) . \end{aligned} \tag{12}$$

$D^{(A,\Gamma)}$ is the covariant derivative with respect to gravitational and gauge field background. The ultraviolet divergence of the integral over M will be cancelled by $G_{\mu\nu}$ and $B_{\mu\nu}$ renormalization leading to the first two lines of eq. (7). The boundary terms should be related to A_μ renormalization. However, first a comment on the relation between $B_{\mu\nu}$ and abelian $F_{\mu\nu}$ is in order. If $\partial M = 0$ the model is invariant with respect to $B \longrightarrow B + d\Lambda$. This invariance is lost in the presence of a boundary. The action S_M now is sensible to pure gauge contributions to B and these cannot be distinguished from $S_{\partial M}$ with abelian A. With respect to A_μ renormalization the terms proportional to the normal derivative pose some problems. They are absent if S = 0, then we get for nonabelian A

$$\begin{aligned} (\beta^A_\lambda)_{abelian} &= -\frac{1}{2\pi}\, D^\mu B_{\mu\lambda} + O(\alpha') \\ \beta^A_\lambda &= -\alpha'\, D^{(A,\Gamma)\mu} F_{\mu\lambda} + O(\alpha'^2) \end{aligned} \tag{13}$$

and for abelian A

$$\beta^A_\lambda = -\alpha' D^\mu \left(F_{\mu\lambda} + \frac{1}{2\pi\alpha'} B_{\mu\lambda} \right) + \cdots \quad . \tag{14}$$

If S $\neq$ 0 one can try to eliminate the normal derivative term by the use of the boundary condition for the background configuration. This is straigthforward for abelian A only

$$\epsilon^{mn} \dot{z}_m \partial_n x^\lambda = -i \left(B^{\lambda\mu} + 2\pi\alpha' F^{\lambda\mu} \right) \dot{x}_\mu \tag{15}$$

leading to

$$\beta^A_\lambda = -\alpha' D^\mu \left(F_{\mu\lambda} + \frac{1}{2\pi\alpha'} B_{\mu\lambda} \right) - \\ - 2\pi\alpha'^2 \left(F_{\alpha\beta} + \frac{1}{2\pi\alpha'} B_{\alpha\beta} \right) S^{\alpha\beta\mu} \left(F_{\mu\lambda} + \frac{1}{2\pi\alpha'} B_{\mu\lambda} \right) + \cdots \quad . \tag{16}$$

For the case of pure gauge field background fields (G = B = ϕ = 0) we find at 2 loop order

$$\beta^A_\lambda = -\alpha' D^{(A)\mu} F_{\mu\lambda} - i\alpha'^2 \left[D^{(A)}_\lambda F_{\mu\nu}, F^{\mu\nu} \right] + O(\alpha'^3) \tag{17}$$

independent on the renormalization scheme and for abelian A at 3 loop order

$$\beta^A_\lambda = -\alpha' \partial^\mu F_{\mu\lambda} - \alpha'^3 4\pi^2 (\partial^\mu F_{\nu\lambda}) F_{\mu\rho} F^{\rho\nu} + O(\alpha'^4) \quad . \tag{18}$$

The equation $\beta^A_\lambda = 0$ agrees at this order of α' with the equation of motion for the generalized Born-Infeld action / 3 /

$$\sqrt{\det(1 + 2\pi\alpha' F)} \quad .$$

4. Gauge field β-function to all orders in α' and dual loop corrections

Using in (11) instead of the usual Neumann function with constant normal derivative a function which satisfies the boundary condition (15) for constant F one finds at lowest order / 7 /

$$\beta^A_\lambda = -\alpha' \partial^\nu F^\mu{}_\lambda \left(1 - 4\pi^2\alpha'^2 F^2 \right)^{-1}_{\mu\nu} \quad . \tag{19}$$

This is equivalent to summing up all diagrams based on the old N and selecting contributions to the first order in the derivative of F. There are claims that (19) is correct for arbitrary abelian F / 8 /. The extension of any kind of these considerations to nonabelian A is nontrivial since the noncommutativity introduces new sources for ultraviolet divergencies which complicate the combinatorics in higher orders.

The trick of ref. / 7 / also works for constant G, B, A, ϕ background and yields for this more general case / 10 /

$$\beta^A_\lambda = \alpha' D^\nu \left(F + \frac{1}{2\pi\alpha'} B\right)_\lambda{}^\mu \left[G - (2\pi\alpha' F + B)^2\right]^{-1}_{\mu\nu} +$$

$$+ 2\pi\alpha'^2 \left(F + \frac{B}{2\pi\alpha'}\right)_{\lambda\nu} S^{\nu\mu\rho} \left[\frac{F + \frac{B}{2\pi\alpha'}}{G - (2\pi\alpha' F + B)^2}\right]_{\mu\rho} . \qquad (20)$$

Assuming as announced in section 2 that the difference between β^A_λ and the corresponding Weyl anomaly coefficient $\bar{\beta}^A_\lambda$ is only due to explicit symmetry breaking by the dilaton coupling one gets / 10 /

$$\bar{\beta}^A_\lambda = \beta^A_\lambda + \frac{\alpha'}{2} D^\nu \phi \left(F + \frac{B}{2\pi\alpha'}\right)_{\lambda\nu} . \qquad (21)$$

The conformal invariance requirement $\bar{\beta}^A_\lambda = 0$ is equivalent to the A_λ equation of motion for the action

$$S^{open}_{eff} = \int d^D x \, e^{\phi/2} \sqrt{\det(G + B + 2\pi\alpha' F)} . \qquad (22)$$

On the other hand $\bar{\beta}^\phi = \bar{\beta}^G = \bar{\beta}^B = 0$ can be deduced as eqs. of motion for

$$S^{closed}_{eff} = \int d^D x \, e^\phi \left[R - 1/3 \, S^2 - (D\phi)^2 - 2 D^2 \phi\right] . \qquad (23)$$

Now the authors of ref. / 10 / make the crucial step to postulate the total effective action as

$$S_{eff} = S^{closed}_{eff} + K \cdot S^{open}_{eff} \qquad (24)$$

with an up to this point free parameter K. While leaving the equation of motion for the gauge field A_μ unchanged the change to S_{eff} modifies the eqs. for the other fields. In particular the gauge field now can act as a source for gravity. This seems to be the most essential achievement of this approach since in the standard picture the gauge field sitting at the boundary cannot influence the closed string field β -functions. Further analysis reveals / 10 / this pure σ -model recipe as the consequence of coupling usual ultraviolet singularities and singularities for a vanishing hole on the string world sheet in the sense of ref. / 9 /.

5. A comment on the 2 dimensional stress tensor in the presence of $B_{\mu\nu}$

Although one expects a boundary contribution containing β^A in the trace anomaly (3) there is still no proof available. As already mentioned the main obstacle is the absence of 2 dimensional g_{ab} in the corresponding coupling term in the action. Just the same situation we find for the coupling to $B_{\mu\nu}$. Since the abelian gauge field via Stokes theorem can be included as a pure gauge part into $B_{\mu\nu}$ it is sufficient to study the $B_{\mu\nu}$ case. To test our technical means we do this first for the closed string $\partial M = 0$. We will find that even in this situation the up to now used treatment via dimensional regularization poses some serious problems.

To have a handle for a $B_{\mu\nu}$ contribution to the stress tensor we use the canonical definition

$$T^{ab} = \frac{1}{4\pi\alpha'} \Big[\left(2\delta^{ma}\delta^{nb} - \delta^{mn}\delta^{ab}\right) G_{\mu\nu}\, \partial_m x^\mu \partial_n x^\nu - i\left(\epsilon^{ma}\delta^{nb} + \epsilon^{mb}\delta^{na} - \delta^{ab}\epsilon^{mn}\right) B_{\mu\nu}\, \partial_m x^\mu \partial_n x^\nu \Big] . \tag{25}$$

We have dropped ϕ and use flat g_{ab}. Eq. (25) agrees with the definition via $\delta/\delta g_{ab}$ and the $B_{\mu\nu}$ dependence vanishes in 2 dimensions since the tensor

$$P^{mnab} = \epsilon^{ma}\delta^{nb} + \epsilon^{mb}\delta^{na} - \delta^{ab}\epsilon^{mn} \tag{26}$$

is annihilated by any factor antisymmetric in m, n. We can however use the dimensional continuation of

$$\epsilon^{ab}\epsilon^{mn} = \sqrt{g}\,(g^{am}g^{bn} - g^{an}g^{bm}) \tag{27}$$

in the evaluation of corresponding Feynman diagrams.

Since T^{ab} of eq. (25) is not derived via a parametric differentiation with respect to g_{ab} one is not sure whether there is a non-renormalization theorem. Of course it would be possible to construct a pure formal recipe along this line of arguments by defining $\frac{\delta\epsilon^{mn}}{\delta g_{ab}}$ in a way to ensure the product rule in (27).

But to be on safer ground we decided to study the one loop renormalization of T^{ab} explicitely. For the general dimension two operator $H_{\mu\nu}(x)\,\partial_m x^\mu \partial_n x^\nu$ we found in arbitrary regularization

$$\begin{aligned}
\langle H_{\mu\nu}(x)\partial_m x^\mu\partial_n x^\nu\rangle_{\tilde{x}} = 2\pi\alpha' N(0)\Big\{ &\partial_m\tilde{x}^\mu\partial_n\tilde{x}^\nu\big[\tfrac{1}{2}D^2H_{\mu\nu} - \tfrac{1}{2}R_\mu{}^\kappa H_{\kappa\nu} \\
&- \tfrac{1}{2}R_\nu{}^\kappa H_{\mu\kappa} - \tfrac{1}{2}H^{\alpha\beta}R_{\mu\nu\alpha\beta} + H^{(\alpha\beta)}S_{\gamma\beta\mu}S^\gamma{}_{\alpha\nu}\big] \\
&- i(\epsilon_{mp}\partial_p\tilde{x}^\mu\partial_n\tilde{x}^\nu - \epsilon_{np}\partial_p\tilde{x}^\mu\partial_m\tilde{x}^\nu)\big[S^{\alpha\beta}{}_\mu D_\beta H_{\alpha\nu} + \tfrac{1}{2}H^{\alpha\beta}D_\nu S_{\alpha\beta\mu}\big] \\
&- \delta_{mn}\Big[\partial_p\tilde{x}^\mu\partial_p\tilde{x}^\nu\big(\tfrac{1}{2}H^{(\alpha\beta)}R_{\alpha\mu\beta\nu} + \tfrac{3}{2}H^{\alpha\beta}S_{\gamma\beta\nu}S^\gamma{}_{\alpha\mu}\big) \\
&- \tfrac{i}{2}\epsilon^{pq}\partial_p\tilde{x}^\mu\partial_q\tilde{x}^\nu H^{(\alpha\beta)}D_\alpha S_{\beta\mu\nu}\Big] \\
&+ \tfrac{i}{2}H^{\alpha\beta}S_{\alpha\beta\mu}\big(\epsilon^{np}D_m\partial_p\tilde{x}^\mu - \epsilon^{mp}D_n\partial_p\tilde{x}^\mu\big)\Big\} + \text{uv finite,}
\end{aligned} \tag{28}$$

where $\tilde{x}$ is a background configuration. Special cases of this formula with summed m, n are contained in ref. /13 /, too.

Applying this formula to T^{ab} and using dimensional regularization $N(0) = -\frac{1}{2\pi}\,\frac{1}{d-2}$ supplemented by relations derived from (27) like

$$P^{mnab}\left(\epsilon_{mp}\partial_p x^\mu \partial_n x^\nu - \epsilon_{np}\partial_p x^\mu \partial_m x^\nu\right) =$$
$$= (d-2)\left(\partial_a x^\mu \partial_b x^\nu + \partial_b x^\mu \partial_a x^\nu - 2\delta_{ab}\,\partial_p x^\mu \partial_p x^\nu\right) \qquad (29)$$

and the convention

$$P^{mnab}\cdot A_{mn} = 0 \qquad (30)$$

for any antisymmetric A_{mn} <u>not</u> containing ϵ tensors we find that the expectation value of T^{ab} is made ultraviolet finite by G and B renormalization only.

$$\begin{aligned}\langle T(G^\circ, B^\circ)_{ab}\rangle_{\tilde{x}} &= T_{ab}(G,B)(\tilde{x}) - \\ &- \tfrac{1}{2\pi}\Big\{\partial_a \tilde{x}^\mu \partial_b \tilde{x}^\nu S^{\alpha\beta}{}_\mu S_{\alpha\beta\nu} - \tfrac{\delta_{ab}}{4}\partial_p \tilde{x}^\mu \partial_p \tilde{x}^\nu\left(R_{\mu\nu} + S^{\alpha\beta}{}_\mu S_{\alpha\beta\nu}\right) \\ &\quad - \tfrac{i}{4}\delta_{ab}\,\epsilon^{pq}\partial_p \tilde{x}^\mu \partial_q \tilde{x}^\nu D^\alpha S_{\alpha\mu\nu} \\ &\quad - \tfrac{1}{4}\Big[D_b\left(\partial_a \tilde{x}^\mu B^{\alpha\beta} S_{\alpha\beta\mu}\right) + D_a\left(\partial_b \tilde{x}^\mu B^{\alpha\beta} S_{\alpha\beta\mu}\right) \\ &\quad - 2\delta_{ab} D_p\left(\partial_p \tilde{x}^\mu B^{\alpha\beta} S_{\alpha\beta\mu}\right)\Big]\Big\} \\ &+ O(\alpha') + \text{nonlocal terms.}\end{aligned} \qquad (31)$$

However, there is a finite renormalization of all components of T^{ab}. For the trace it is welcome since

$$\begin{aligned}\langle T(G^\circ, B^\circ)_{aa}\rangle_{\tilde{x}} = \tfrac{1}{4\pi}\Big[&\partial_a \tilde{x}^\mu \partial_a \tilde{x}^\nu\left(R_{\mu\nu} - S^{\alpha\beta}{}_\mu S_{\alpha\beta\nu}\right) \\ &+ i\epsilon^{ab}\partial_a \tilde{x}^\mu \partial_b \tilde{x}^\nu D^\alpha S_{\alpha\mu\nu} \\ &- D_p\left(\partial_p \tilde{x}^\mu B^{\alpha\beta} S_{\alpha\beta\mu}\right)\Big]\end{aligned} \qquad (32)$$

fits into (3). The finite renormalization of the off diagonal elements of T^{ab} in the case $S \neq 0$ with its related interpretation as a violation of 2 dimensional general covariance should be an artifact of dimensional regularization. Just the same conclusion with an $S \cdot S$ term but without the total derivative part has been reached in ref.

/ 19 / by an analysis of the conformal operator product expansion for two T's. Obviously, further investigations of the antisymmetric tensor coupling in a renormalization scheme independent way are necessary to exclude an unwanted anomaly in the off diagonal elements of T^{ab}.

References

/ 1 / M. Green and J. H. Schwarz, Phys. Lett. 149 B (1984) 117

/ 2 / B. Zwiebach, Phys. Lett. 156 B (1985) 315

/ 3 / E. S. Fradkin and A. A. Tseytlin, Phys. Lett. 163 B (1985) 123

/ 4 / A. A. Tseytlin, Phys. Lett. 176 B (1986) 92

/ 5 / H. Dorn and H.-J. Otto, Proc. Symp. Ahrenshoop 86, Berlin - Zeuthen, IfH PHE 87-

/ 6 / H. Dorn and H.-J. Otto, Z. Phys. C 32 (1986) 599

/ 7 / A. Abouelsaood, C. G. Callan, C. R. Nappi and S. A. Yost, Princeton prepr. (1986)

/ 8 / E. Bergshoeff, E. Sezgin, C. N. Pope and P. K. Townsend, Trieste prepr. (1986)
C. N. Pope, talk at this workshop

/ 9 / W. Fischler and L. Susskind, Phys. Lett. 173 B (1986) 262

/10 / C. G. Callan, C. Lovelace, C. R. Nappi and S. A. Yost, Princeton prepr. (1986)

/11 / A. A. Tseytlin, Phys. Lett. 178 B (1986) 34

/12 / G. Curci and G. Paffuti, CERN prepr. TH 4558/86
G. M. Shore, Bern prepr. BUTP-86/16

/13 / A. A. Tseytlin, Lebedev prepr. 342 (1986)

/14 / D. Friedan, Phys. Rev. Lett. 45 (1980) 1057 and Ann. Phys. 163 (1985) 318

/15 / C. G. Callan, E. Martinec, M. J. Perry and D. Friedan, Nucl. Phys. B 262 (1985) 593

/16 / E. Braaten, T. L. Curtright and C. K. Zachos, Nucl. Phys. B 260 (1985) 630

/17 / B. Fridling and A. E. M. van de Ven, Nucl. Phys. B 268 (1986) 719

/18 / M. T. Grisaru, A. E. M. van de Ven and D. Zanon, Nucl. Phys. B 277 (1986) 388, 409
M. T. Grisaru, D. I. Kazakov and D. Zanon, Harvard Univ. prepr. HUTP-86 A 076

/19 / T. Banks, D. Nemeschansky and A. Sen, Nucl. Phys. B 277 (1986) 67

Discussion session on part II: Non-linear σ-models

A number of questions asked repeatedly during the individual lectures were collected and presented at the discussion session. They concerned the topics

i) Off-shell IR-problem;

ii) Covariance properties of Green's functions and observable quantities;

iii) Renormalizability;

iv) Conformal invariance.

The written answers to these questions are not those of any individual speaker, but represent the general opinion emerging during the discussion.

To i): The free part of the action of the non-linear σ-model describes a massless scalar field. In space-time dimensions $d \leq 2$ a given background field configuration is unstable against long-wavelength fluctuations of this massless field in view of the IR-singularity of its 2-point function $\int \frac{d^2k}{k^2} e^{ikx}$. Therefore a mass term has to be introduced which stabilizes a given classical background configuration. This automatically breaks the Ward (Slavnov) identity furnishing the independence of the Green's functions of the background configuration (compare Becchi's lecture). This breaking is considered to be soft (Stelle's lecture).

Question: Is the mass term really soft or is it potentially harmful?

Answer: Concerning the renormalization procedure the effect of the mass term in the UV can be separated by power-counting. This, however, does not mean that the theory has a finite zero-mass limit and if so that the latter is correct. Experience with the $O(N)$-model (David) suggests the following situation. If the target manifold is a coset-space and if the mass term has a well-defined covariance with respect to the isometries of the space it should be possible to extend the result of David saying that Green's functions which are invariant under the isometries have a finite zero-mass limit. This ought to be connected with the fact that the breakdown of the isometry artificially introduced by the mass term is averaged to zero in the case of invariant Green's functions. Nothing is known in the general case of non-standard models (no coset-space). Even if the IR-limit exists in perturbation theory the result may be incorrect as the case $d = 1$ shows. It may be that the perturbative approach does not adequately reproduce the large field fluctuations at long wavelength.

To ii): It is possible in the construction of non-linear σ-models to use covariant (normal geodesic) coordinates for which the covariance of the action and the counter terms under coordinate transformations is preserved using dimensional regularization with minimal subtractions.

Question: What are the covariance properties of connected Green's functions?

Answer: The generating functional for connected Green's functions (free energy) depending on linear field sources has no simple covariance property under coordinate transformations. However, in principle, covariance could be rescued for the 'effective action' defined as the Legendre transform of the connected functional. One should notice that in the case of coset-spaces the connected functional is constrained by the Ward identity representing the action of the isometry group on the coset-space and that this is sufficient to define the renormalized theory.

Question: How can one understand the covariance of the 'effective action' using dimensional renormalization?

Answer: A simple argument in favour of the covariance of dimensional renormalization with minimal subtraction could be based on the fact that the covariant construction of the dimensionally regularized theory in terms of the normal geodesic coordinates with respect to the bare metric tensor is valid in any integer space-time dimension. This ceases to be true if a Wess-Zumino term is added to the σ-model action.

Question: Do Green's functions for coordinates in the target manifold make any sense or should one restrict oneself to the discussion of observables and what are relevant observables ?

Answer: Everybody agrees that there is no observable meaning to coordinates of a manifold. On the other hand the Green's functions for coordinates may contain relevant information about observable quantities such as the particle spectrum (masses etc.). The situation may be viewed in analogy to gauge theories where there seems to be no doubt about the usefulness of the Green's functions of gauge potentials having a similar status as coordinates. On the other hand it would be definitely desirable to construct Green's functions for observables like the energy-momentum tensor. This would in particular allow to extract information about the current algebra of the energy-momentum tensor relating to the conformal properties (Virasoro algebra) of the theory.

To iii): There is ample literature about the renormalization of non-linear σ-models in two space-time dimensions in particular for special types of manifolds (Kähler etc.). Of special interest are supersymmetric extensions.

Question: Are non-linear σ-models in two space-time dimensions renormalizable theories?

Answer: Everybody agrees that the general non-linear σ-model for an arbitrary Riemannian manifold is not renormalizable in the conventional sense, because the action is unstable against arbitrary deformations of the metric. This instability leads to infinitely many parameters of the renormalized theory. On the other hand there is also agreement that renormalizability could be proved for coset spaces except for the cases where there are antisymmetric couplings. This could be done either using an invariant regularization (e.g. dimensional) or by discussing in detail the cohomology of the isometry group. It is suggested that some special G-structures – that is Riemannian manifolds whose holonomy group is G, a subgroup of the orthogonal group – could perhaps lead to renormalizable models with a finite number of parameters. This however does not seem to be true for Kähler manifolds where

G is a unitary group. Another possible such case is that of hyper-Kähler models having the scalar content of $N = 4$ supersymmetric models. In any case, nothing is known for supersymmetric models, because dimensional renormalization is not an invariant scheme for such theories.
Assuming that the Wilson renormalization group preserves global topological constraints and that fixed points correspond to renormalizable models one might consider 2-dimensional target manifolds. In this case the only topological invariant is the genus of the manifold. It is suggestive to speculate that the fixed points yield algebraic surfaces, even if some arguments (Lott) seem to indicate an instability of the constant curvature configurations (e.g. spheres) against the formation of corners.

To iv): In connection with string theories a particular emphasis is put on the construction of non-linear σ-models with vanishing β-function (conformal invariance).

Question: Is the vanishing of the β-function ('finiteness') a possible handle on the renormalization problem for manifolds without isometries, e.g. hyper-Kähler models with a Wess-Zumino term?

Answer: The requirement of conformal invariance seems to restrict rather drastically the ambiguities involved in defining these theories (Flume), to the extent that explicit formulae can be given for the 4-point functions of primary fields. However, the construction of the Virasoro algebra for the energy-momentum tensor has not yet been achieved in renormalized perturbation theory.

Part III

Cohomological and Geometrical Methods, Relation to String Theory

A free string moving in a classical background manifold can be described by the equations of motion for a quantized non-linear σ-model in two-dimensional space-time with this manifold as target space.

Cohomological and geometrical methods play an important rôle in renormalization theory, in non-linear σ-models and in string theory. It is tried to point out possible links between these subjects.

Remarks on Slavnov Symmetries

R. STORA

LAPP, Annecy le Vieux, France
and
Theory Division, CERN
Geneva, Switzerland

Slavnov symmetries have been discovered some ten years ago [1] within the framework of perturbative Yang-Mills type gauge theories. They were found as non-linear local field transformations, depending on an anticommuting parameter, which leave the Faddeev Popov-gauge fixed Yang-Mills action invariant. They turned out to be the ideal tool to perform the perturbative renormalization of Yang-Mills theories, taking full advantage of *locality* and power counting.

It was soon realized [1] [2] that the Slavnov symmetry could be cast into the form of a cohomological statement related to a differential algebra which is a trivial extension of the cohomology algebra of the gauge Lie algebra.

This class of algorithms was extended to the analysis of a number of similar situations:

- other gauge Lie algebras including diffeomorphisms, and their local anomalies;
- graded Lie algebras occurring e.g. in globally as well as locally supersymmetric theories;
- more general differential algebras [3];
- constrained systems in Hamiltonian dynamics [4].

Similar constructions have also proved useful in statistical mechanics, and, in particular in the method of stochastic quantization [5].

The subject has been actively revived recently outside the small group of adepts in connection with string theory, mostly, however, in the form close to the Hamiltonian version analyzed by the Lebedev group [4], in which both locality and geometry fail to serve as leading principles. (See however [6]).

Let us finally mention the applicability of Slavnov type algorithms to express coordinate independence of σ-models [7] one of the main topics covered during this meeting.

Given this list of examples, one may wonder what features they share in common, besides being cohomology theories.

One answer is the following:

One starts from a (local) classical action $S(\underline{\phi})$ whose arguments are fields, some of which will be quantized, whereas some others will remain classical sources. Consider the situation where $S(\phi)$ admits infinitesimal zero modes

$$\delta\phi(x) = P_i(\phi)\delta\lambda^i(x) \tag{1}$$

where the P_i's are ϕ dependent linear operators and the $\delta\lambda^i$'s infinitesimal parameters, i.e.

$$\int dx \frac{\delta S}{\delta\phi^\alpha(x)}\delta\phi^\alpha(x) \equiv 0 \tag{2}$$

for $\delta\phi$ given by (1). In gauge situations the $\delta\lambda^i$'s are fields and the $Pi's$ are differential operators which depend locally on ϕ.

If these zero modes are integrable into " leaves" in field space, one may consider the λ^i's as parameters along these submanifolds, and δ as differentiation along these submanifolds: locally in field space, one has a parametrization

$$\begin{aligned} \phi(x) &= \Phi(\phi_o, \lambda)(x) \\ \delta_\lambda\phi &= \int \frac{\delta\Phi}{\delta\lambda(y)}\delta\lambda(y) = Q_i(\phi)\delta\lambda^i \end{aligned} \tag{3}$$

where Q_i are (not necessarily local) linear operators. So, we may define

$$\begin{aligned} \omega^i &= \delta\lambda^i \\ s &= \delta \to s^2 = 0 \end{aligned} \tag{4}$$

so that $S(\phi)$ is s-invariant, with

$$\begin{aligned} s\phi &= Q_i(\phi)\omega^i \\ s\omega^i &= 0; \end{aligned} \tag{5}$$

of course $s^2 = 0$.

Conversely given (5) such that $s^2 = 0$ Frobenius formally guarantees that the infinitesimal zero modes are integrable. The locality question is not under control and is sensitive to the choice of parametrization (3). For instance, in the Yang-Mills case let us parametrize the Yang-Mills field a according to

$$a = g^{-1}\overset{\circ}{a}g + g{-}1dg \tag{6}$$

with g in the gauge group, $\overset{\circ}{a}$ subject to a gauge condition. We have

$$\delta_g a = D_a(g^{-1}\delta g) = -d(g^{-1}\delta g) - [a, g^{-1}\delta g] \tag{7}$$

(assuming $\delta g d + d\delta g = 0$). δg plays here the role of $\delta\lambda$. Trying to eliminate g in terms of a, in (7) would in general result into non locality. However, we know that, defining

$$\begin{aligned} \omega &= g^{-1}\delta g \\ \delta &= s \end{aligned}$$

we have

$$\begin{aligned} &sa = -d\omega - [a, \omega] \\ &s\omega = -\frac{1}{2}[\omega, \omega] \\ &s^2 = 0 \end{aligned} \tag{8}$$

which are the well known local formulae for the cohomology of the gauge Lie algebra.

When zero modes affect quantized fields, the Faddeev-Popov procedure allows to extend the initial differential algebra by introducing the antighost field $\bar{\omega}$ and the Lagrange multiplier b in such a way that

$$\begin{aligned} s\bar{\omega} &= -b \\ sb &= 0 \end{aligned} \tag{9}$$

leaves the gauge fixed action

$$S_{gf} = (b, g) + (\bar{\omega}, sg) \tag{10}$$

invariant, whatever gauge functions g have been chosen. It is to be noted that the geometrical natures of $\bar{\omega}, b$ essentially depend on the choice of the gauge functions g.

When the zero modes do not give rise to an integrable situation no general theory is known at the moment: this is the problem of auxiliary fields, namely, given a non-integrable situation, can it be viewed as a restriction of an integrable one?

It may happen that, due to a "bad" choice of a gauge function and or imposing locality, zero modes appear in the ghost sectors [6] [8]. This poses a slightly more general problem namely, given s with $s^2 = 0$ acting on some set of fields ϕ, and given ϕ-dependent zero modes $\Phi_i(\phi)$:

$$\delta\phi = \sum_i c_i \Phi_i(\phi) \tag{11}$$

can one find $\mathcal{S} c_i$ such that together with

$$\mathcal{S}\phi = s\phi - \sum c_i \Phi_i(\phi) \tag{12}$$

one achieves $\mathcal{S}^2 = 0$.

It is fair to say that no general scheme has been found so far which is able to deal with the quantization of degenerate actions as stated above; only classes of examples have been collected, of which we have quoted representatives.

ACKNOWLEDGEMENTS

These remarks constitute by no means a fair review of the work which has been accomplished and contributes to the understanding of Slavnov symmetries. They rather try to extrapolate from the experience which has been gained by collecting and working out examples towards the formulation of some open problems. The author wishes to thank many colleagues for sharing their views concerning this intriguing area, and, in particular, C. Becchi, L. Baulieu, L. Bonora, P. Cotta-Ramusino, O. Piguet, K. Sibold. Also, it is a pleasure to thank the organizers of this meeting for the opportunity they gave us to meet our colleagues and hear about recent developments.

REFERENCES

[1] C. Becchi, A. Rouet, R. Stora, Ann.Phys. **98** (1976) 287.

[2] C. Becchi, A. Rouet, R. Stora, in: *Renormalization Theory* Erice 1975; G. Velo, A.S. Wightman, Ed. NATO ASI Series Vol.C23, Reidel 1976; C. Becchi, A. Rouet, R. Stora in: *Field Theory Quantization and Statistical Mechanics* Tirabegin ed. Reidel 1981.

[3] S. Boukraa: *The BRS Algebra of a free differential algebra*, *Nucl. Phys.* **B** (to appear).

[4] M. Henneaux: Phys.Rep. **126** (1985) 1.

[5] Zinn-Justin, *Nucl. Phys.* **B 275** (1986) 135[FS17]

[6] L. Baulieu, C. Becchi, R. Stora, *Phys. Lett.* **180 B** (1986) 55
C. Becchi: *On the covariant quantization of the free string: the conformal structure Nucl. Phys.* **B** (to appear); L. Baulieu, M. Bellon, *Beltrami parametrization and string theory,* LPTHE 87-39;
L. Baulieu, M. Bellon, R. Grimm: *Beltrami parametrization for superstrings*, LPTHE 87-43;
L. Baulieu, B. Grossmann, R. Stora *Phys. Lett.* **180 B** (1986) 95;
J.P. Ader, J.C. Wallet *Phys. Lett.* **192 B** (1987) 103.

[7] See e.g. C. Becchi's talk at this meeting.

[8] L. Baulieu, J. Thierry-Mieg *Nucl. Phys.* **B 228** (1983) 259.

SUPERSYMMETRIC PROPERTIES OF FIELD THEORIES IN 10-D

L. Bonora [1][2], M. Bregola [4], K. Lechner [3],
P. Pasti [2][3], M. Tonin [2][3]

(1) CERN, TH Division

(2) Dipartimento di Fisica dell'Università di Padova, Via Marzolo 8, Padova.

(3) INFN, Sezione di Padova

(4) Dipartimento di Fisica dell'Università di Ferrara, Via Paradiso 12, e INFN, Sezione di Bologna.

* * *

A field theory which represents the low energy limit of a string theory is perhaps the most important piece of information that we can extract from that string theory. The effective field theory can be obtained by direct string amplitude calculations; or else we can start from the field theory content, which is unambiguosly defined by the zero mass excitations of the string, and follow at present two methods. The first consists in writing down a sigma model of the string interacting with background fields corresponding exactly to the zero mass spectrum of the string: the requirement of conformal invariance, through the vanishing of the beta functions, will provide us the equations of motions of the background fields. In the second approach one simply takes from the string theory the field content and with these fields tries to construct a field theory. If this field theory turns out to be consistent and, especially, if it enjoys some kind of uniqueness it, hopefully, will represent the low energy limit of the string theory.

In this talk we will be concerned with the second approach. It is well known that the zero modes of the heterotic string and type I superstring correspond to the field content of a supersymmetric Yang-Mills (SYM) theory coupled to supergravity (SUGRA) in ten dimensions. In other words, beside the graviton, the dilaton, the "two-index photon" and the gauge potentials, we have the gravitino, the dilatino and the gauge superpartners.

Of course constructing a field theory with such a field content is interesting in itself, even without reference to string theory.

However the number of theories one could construct would be enormous, due to the different choices of the gauge group , were it not for the requirement of chiral anomaly cancellation. As is well known [1] chiral anomalies cancel in 10-D SYM+SUGRA theories provided that the gauge group is $E_8 \times E_8$ or SO(32), and provided a two form field B and a three form field H (its "curvature") are defined in such a way as to satisfy the equation

$$H = dB + c_1 \omega_{3Y} - c_2 \omega_{3L} \tag{1}$$

where ω_{3Y} (ω_{3L}) is the gauge (Lorentz) Chern-Simons form i.e.

$\omega_{3Y} = \mathrm{Tr}(AdA + \frac{2}{3}AAA)$, $\omega_{3L} = \mathrm{Tr}(\omega d\omega + \frac{2}{3}\omega\omega\omega)$, $A(\omega)$ being the gauge (Lorentz) connection; c_1 and c_2 are constants, which must assume particular values for the anomaly to cancel (for example, for SO(32), $c_1 = c_2 = 1$). In eq. (1) B is identified with the two-index photon. This identification raises several problems which are in part addressed here and in part in a parallel talk [2]. Here we study the problem of implementing supersymmetry in the presence of condition (1). This problem arose historically in the following way. Chapline and Manton [3] succeeded a few years ago in writing down a lagrangian of a SYM theory coupled to SUGRA in 10-D. They assumed a condition similar to (1) except that $c_2 = 0$. As a consequence their theory is plagued by chiral anomalies. Green and Schwarz, through eq. (1) were able to cancel anomalies, however in this way they destroyed manifest supersymmetry in the Chapline-

-Manton lagrangian.

There have been a few attempts [4] to restore supersymmetry by suitably modifying the Chapline-Manton lagrangian, but this approach appears rather unwieldy. Another, more promising approach is based on supermanifold techniques [5,6,7]. Supermanifold techniques have the disadvantage that, at least in ten dimensions, they cannot give us a Lagrangian but only the equations of motion, but of course they have the advantage of involving superfields instead of component fields and of resorting to a "supergeometry" which is very powerful (although probably not still understood). The supermanifold approach is the one we have used to solve the problem of implementing supersymmetry in the presence of condition (1).

Let us consider a supermanifold with coordinates $z^{\mu} = \{x^{u}, \theta^{\mu}\}$, m=1,...,10, μ=1,...,16. We will use throughout mostly flat-indexed objects. Flat indices are denoted by the first letters of the alphabet: ex. A={a,α}, a=1,...,10, α=1,...,16. In order to pass from world indices to flat ones we introduce the supervielbeins e^{A}_{M} (z) and the corresponding superforms $e^{A} = dz^{M} e^{A}_{M}(z)$. Next we mimic usual differential geometry and introduce a Lorentz superconnection $\omega_{A}{}^{B}$ and a Lie algebra-valued superconnection A together with the torsion T^{A}, the curvatures $R_{A}{}^{B}$ and F given by

$$
\begin{aligned}
T^{A} &= \Delta e^{A} \\
F &= dA + AA \\
R_{A}{}^{B} &= d\omega_{A}{}^{B} + \omega_{A}{}^{C}\omega_{C}{}^{B}
\end{aligned}
\tag{2}
$$

Here $d = dz^{M}\frac{\partial}{\partial z^{M}}$ is the total differential in the superspace, while $\Delta = e^{A}\Delta_{A}$ is the covariant differential w.r.t. both gauge and Lorentz indices. Moreover we introduce a 3-superform H and a 2-superform B and write the analogue of eq. (1) where the fields are replaced by the corresponding superfields. Henceforth eq. (1) will be regarded as a superfield equation. Corresponding to eqs. (1) and (2) we have the identities

$$dH = c_1 Tr(FF) - c_2 Tr(RR) \tag{3}$$

$$\Delta T^A = e^B R_B{}^A \tag{4}$$

$$\Delta R_A{}^B = 0 \tag{5}$$

$$\Delta F = 0 \tag{6}$$

We use the basis of supervielbeins $\{e^A\}$ to define the components of a given superform. For instance $T^A = \frac{1}{2} e^B e^C T_{CB}{}^A$.

In order to extract, from this very general geometric setting, the equations of motion, the procedure consists in reducing the number of degrees of freedom by imposing constraints on the components of T^A, F and H. The Bianchi identities (3)(4) and (6) will produce the equations of motion for the surviving degrees of freedom. Eq. (5) does not give additional information.

Following [7] we postulate the following constraints

$$T^a_{\alpha\beta} = 2\Gamma^a_{\alpha\beta} \; ; \quad T_{\alpha\beta}{}^\gamma = 0 = T_{\alpha c}{}^b \; ; \quad T_{a\beta}{}^\gamma = \Gamma_{a\beta\delta}\psi^{\delta\gamma} \tag{7}$$

$$F_{\alpha\beta} = 0 \tag{8}$$

$$H_{\alpha\beta\gamma} = 0 \quad ; \quad H_{a\beta\gamma} = \phi\Gamma_{a\beta\gamma} \tag{9}$$

The constraints (9) will be modified in the following by terms proportional to c_2 (see below).

Here we use two sets of Γ-matrices, with upper and lower indices, respectively. They satisfy the condition

$$\Gamma^a_{\alpha\beta}\Gamma^{b\beta\gamma} + \Gamma^b_{\alpha\beta}\Gamma^{a\beta\gamma} = 2\eta^{ab}\delta^\gamma_\alpha$$

Moreover $\Gamma^{a_1 \dots a_k}$ stands for the antisymmetrized product of k Γ-matrices normalized to unit weight.

With the above constraints it is possible to solve completely the T-Bianchi identities, eq. (4). One obtains for example,

$$D_\beta \psi^{\beta\gamma} = 0 \ ; \quad \psi^{\alpha\beta} = -\frac{1}{24} T_{abc} (\Gamma^{abc})^{\alpha\beta} \ ; \quad \psi^{\alpha\beta} = -\psi^{\beta\alpha}$$

$$T_{abc} = T_{[abc]} \ ; \quad D_a T^{abc} = 0 \ ; \quad T_{ab}{}^{\gamma} = \frac{1}{4} (\Gamma^c)^{\gamma\delta} D_\delta T_{cab} + 2 (\Gamma_{[a})^{\gamma\delta} L_{\delta b]} \tag{10}$$

where $L_{\alpha b} \equiv \Gamma^c{}_{\alpha\beta} T^\beta_{cb}$

The constraints also imply

$$R_{\alpha\beta}{}^{b_1 b_2} = \frac{1}{6} (\Gamma^{b_1 b_2 c_1 c_2 c_3})_{\alpha\beta} T_{c_1 c_2 c_3} + 3 \Gamma_{c\alpha\beta} T^{c b_1 b_2} \tag{11}$$

and similar expressions for the other components of the curvature.

The F-Bianchi identities, eq. (6), give

$$F_{a\alpha} = \Gamma_{a\alpha\beta} \chi^\beta \ , \quad F_{ab} = \frac{1}{4} (\Gamma_{ab})_\alpha{}^\beta D_\beta \chi^\alpha \tag{12}$$

which leads to the field equations in the pure YM sectors. The H--Bianchi identities, eq. (3) have been solved for the particular case $c_2=0$. In this way the equations of motion of the Chapline--Manton model have been recovered. Explicit formulas can be found for example in ref. [7].

It is worth recalling at this point that the identification with the physical fields is obtained by associating the dilaton to the field Φ of eq. (9) at $\theta=0$, the dilatino to $D_\alpha \Phi$ for $\theta=0$, while graviton and gravitino are associated with the space-time-like vielbein e_m^a and spinor-like vielbein e_m^α respectively. The remaining identifications are the obvious ones.

Before extending the previous analysis to the case $c_2 \neq 0$ we have to introduce some notational conventions. First we decompose any n--superform ψ into $\sum_{p+q=n} \psi_{p,q}$ where $\psi_{p,q}$ is an n-superform homogeneous of degree p in the space-time-like vielbeins e^a and of degree q in the spinor-like vielbeins e^α. We do the same with the exterior differential operator d (when applied to scalar forms). To this end it is convenient to regard it as an operator sending (p,q)-superforms into a sum of (p',q')-superforms and split it into homogeneous pieces according to the degree $(r,s)=(p'-p,q'-q)$. Therefore

$$d = \bar{d} + D + T + \tau \tag{13}$$

where $\bar{d}, D, T, \tau$ have degree (1,0), (0,1),(-1,2) and (2,-1) respectively . More precisely $\bar{d}=e^a D_a + T_{(1,0)}$, $D=e^\alpha D_\alpha + T_{(0,1)}$, $T=T_{(-1,2)}, \tau=T_{(2,-1)}$: D_a and D_α are the covariant flat space-time and spinor derivatives acting only on the components, while $T_{(r,s)}$ acts only on the vielbein basis. In particular $Te^a=\Gamma^a{}_{\alpha\beta}\, e^\alpha e^\beta$, $Te^\alpha=0$.

From the nilpotency of d, $d^2=0$, a set of useful relations among $\bar{d}$,D,T and τ follow. In particular:

$$T^2 = 0 \quad , \quad DT+TD=0 \quad , \quad \bar{d}T+T\bar{d}+D^2=0 \quad .$$

Now we are ready to turn to the discussion of the general case, $c_2 \neq 0$. The difficulty in this case lies on the fact that, if we set $Q=Tr(R,R)=\sum_{i+j=4} Q_{i,j}$, $Q_{0,4}$ and $Q_{1,3}$ are non-vanishing unlike the corresponding superforms in Tr(F,F) (which vanish due to the constraints 8). As a consequence the constraints (9) are untenable. There exists however the opportunity to use the results of the case $c_2=0$ to solve the case $c_2 \neq 0$, by relying on the following remarkable property of Tr(RR). [9]

Lemma

$$Tr(RR) = dX + K \tag{14}$$

where X is a gauge and Lorentz invariant 3-superform and K is a gauge and Lorentz invariant 4-superform such that $K_{1,3}=0=K_{0,4}$ and, moreover $K_{2,2}$ has the same structure as $Tr(F,F)_{2,2}$.

The proof goes as follows. From the identity $dTr(R,R)=0$ we write down the equations

$$T\,Q_{2,2} + D\,Q_{1,3} + \bar{d}\,Q_{0,4} = 0 \tag{15a}$$

$$T\,Q_{1,3} + D\,Q_{0,4} = 0 \tag{15b}$$

$$T\,Q_{0,4} = 0 \tag{15c}$$

In other words, eq. (15c) means that $Q_{0,4}$ is a cocycle of the coboundary operator T. It is easy to show that it is indeed a coboundary

$$Q_{0,4} = T\,Y_{1,2} \tag{16}$$

where $(Y_{1,2})_{a\alpha\beta} = R_{\alpha\beta}{}^{c_1 c_2}\,T_{a c_1 c_2}$. Inserting eq. (16) into (15b) we obtain

$$T(Q_{1,3} - D\,Y_{1,2}) = 0 \tag{17}$$

The cocycle $Q_{1,3}-DY_{1,2}$ is not a coboundary because of the presence of a 1440 irreducible representation of SO(1,9). However one can prove that, if we introduce the three-superform $Z_{1,2}$ such that

$$(Z_{1,2})_{\alpha\beta\gamma} = \frac{1}{6}(\Gamma_a{}^{b_1 b_2 b_3 b_4})_{\beta\gamma} D_{b_1} T_{b_2 b_3 b_4} \tag{18}$$

then $TZ_{1,2}=0$ and

$$Q_{1,3} - DY_{1,2} = DZ_{1,2} + TX_{2,1} \tag{19}$$

Then setting $X_{1,2}=Y_{1,2}+Z_{1,2}$ we have

$$Q_{1,3} = DX_{1,2} + TX_{2,1} \tag{20}$$

$$Q_{0,4} = TX_{1,2} \tag{21}$$

Now inserting eq. (20) into eq. (16a) one finds

$$T(Q_{2,2} - DX_{2,1} - \bar{d}X_{1,2}) = 0 \tag{22}$$

Again this T-cocycle is non-trivial, however one can write in complete generality:

$$Q_{2,2} = DX_{2,1} + \bar{d}X_{1,2} + TX_{3,0} + K_{2,2} \tag{23}$$

where

$$(K_{2,2})_{\alpha\beta,ab} = (\Gamma_{[a})_{\alpha\gamma} W^{\gamma\delta} (\Gamma_{b]})_{\delta\beta} \tag{24}$$

Since $$(\mathrm{Tr}(FF)_{2,2})_{ab\alpha\beta} = (\Gamma_{[a})_{\alpha\gamma}\, \mathrm{Tr}(\chi^{\gamma}\chi^{\delta})\, (\Gamma_{b]})_{\delta\beta}$$
we see that $K_{2,2}$ has the same group theoretical structure. Defining $X=X_{1,2} + X_{2,1} + X_{3,0}$ and setting $K=Q-dX$, we see that the lemma is proved.

Now set

$$\hat{H} = H + c_2 X \tag{25}$$

From eq. (3) one gets the identity

$$d\hat{H} = c_1 \mathrm{Tr}(FF) - c_2 K \tag{26}$$

which is formally identical to the case $c_2=0$ since K and Tr(F,F) have the same structure. Therefore by imposing the new constraints

$$\hat{H}_{\alpha\beta\gamma} = 0\ ; \quad \hat{H}_{a\beta\gamma} = \Gamma_{a\beta\gamma}\, \phi \tag{27}$$

which are formally the same constraints as in the case $c_2=0$, we can solve all the Bianchi identities. Moreover we can use the equations already found for the case $c_2 \neq 0$ and obtain the corresponding equations for $c_2=0$, by simply replacing everywhere H with $\hat{H}$ and Tr(FF) with $c_1\mathrm{Tr}(FF)-c_2K$. All the quantities $X_{2,1}$, $X_{3,0}$, $K_{2,2}$, $K_{3,1}$, $K_{3,1}$, $K_{4,0}$ as well as the equations of motion can be explicitly calculated. The complete results are reported elsewhere [10]. Here we would like to make a few remarks.

One of the equations obtained is the following:

$$\hat{H}_{abc} = -\frac{3}{2}\,\phi\, T_{abc} + \frac{c_1}{4}\mathrm{Tr}(\chi^{\alpha}\chi^{\beta}) - c_2 (\Gamma_{abc})_{\alpha\beta} W^{\alpha\beta} \tag{28}$$

T_{abc} appears in all the equations as an auxiliary superfield. In particular in eq. (28) it appears also in a highly non linear way in the explicit expression of $W^{\alpha\beta}$. If we want to eliminate the dependence on T_{abc} we may solve recursively eq. (28), obtaining an infinite series in R,L and H. This is what the approximate solutions already existing in the literature suggested.

It is clear that we are helped to think the field theory we have found has something to do with string theory, if it is characterized by some kind of uniqueness, or, at least, it is not a member of an infinite set of inequivalent field theories (equivalence here is meant up to field redefinition). A critical point in this regard is the choice of the constraints (7) (8) and (9). Other constraints have been proposed [5], but they are shown to be equivalent to these.

So one might hope that proving uniqueness, after all, is not such a forbidding task. However, regarded in complete generality, the problem is daunting, the number of irreducible components of the fields involved being too high. We need some restrictions. Fortunately a few restrictions are provided by the anomaly problem, by which we mean finding non trivial solutions of the Wess-Zumino consistency conditions which reduce to the usual (non supersymmetric) ones when we take $\theta=0$ in the superconnections. When in the usual case a non trivial solution exists, while in the superfield formulation it does not, we assume this as an inconsistency of the theory. We transform this into a selective criterion for consistent superfield formulations. It is rather easy to show with the method of ref. [11], that in the pure gauge sector the constraints (8) is a necessary and sufficient condition for such non trivial solution to exist. It is also easy to realize that the existence of a solution depends on constraints on the relevant curvature. In the pure SUGRA sector the situation is more complicated than in the pure SYM sector because $R_{\alpha\beta a}{}^{b}$ contains many irreducible components w.r.t. SO(1,9) [Eq. (11) is a very particular representation of $R_{\alpha\beta a}{}^{b}$, which is due to the constraints (7)]. Nevertheless the above criterion seems to impose extremely sharp restrictions. Work in this direction is in progress.

The last remark concerns ghosts. The equations of motion we find [10], unambiguosly contain ghosts. For example the equation for the Ricci tensor takes the form

$$R_{ab} \sim \Box R_{ab} + \cdots$$

which unveils the presence of a ghost which propagates at the Planck mass. Such ghosts are likely to be eliminated by non-local field redefinitions [12]. However one can argue whether such a procedure is a correct one. Perhaps these ghosts are a consequence of the genuine stringy nature of the field theory we are considering. From this point of view they may be far from unappealing.

References

[1] M. Green and J. Schwarz, Phys. Lett. 149B, 117 (1984).

[2] L. Bonora, P. Cotta-Ramusino, and M. Rinaldi these proceedings.

[3] G.F. Chapline and N.S. Manton, Phys. Lett. 120B, 105 (1983).

[4] L. Romans and N. Warner, Caltech preprint CAL-68-1291, 1985. S.K. Han, J. Kim, I. Koh and Y. Tanii, Phys. Rev. D34 (1986), 533.

[5] B.E.W. Nilsson, Nucl. Phys. 188 (1981), 176. B.E.W. Nilsson and A.K. Tollsten, Phys. Lett. 169B (1986), 369; Phys. Lett. 171B (1986), 212.

[6] E. Witten, Nucl. Phys. B266 (1986), 245.

[7] J. Atik, A. Dhar and B. Ratra, Phys. Rev. D33 (1986) 2824.

[8] S. McDowell and M. Rakowski, Yale preprint YTP, 86-01.

[9] L. Bonora, P. Pasti and M. Tonin, DFPD 21/86, to appear in Phys. Lett. B.

[10] L. Bonora, M. Bregola, K. Lechner, P. Pasti and M. Tonin, in preparation.

[11] L. Bonora, P. Pasti and M. Tonin, Nucl. Phys. B286 (1987), 150.

[12] S. Deser and A.N. Redlich, Phys. Lett. 176B, 350 (1986).

GENERALIZED WESS-ZUMINO TERMS

L. Bonora (*)

Theory division, CERN, CH-1211 Geneve 23

P. Cotta-Ramusino (**)

Dipartimento di Fisica dell'Università di Milano

and Istituto Nazionale di Fisica Nucleare, Sezione di Milano

Via Celoria 16, 20133 Milano

M. Rinaldi

International School for Advanced Studies

Strada Costiera 11, 34100 Trieste

A possible way to study string theories is to write down sigma-models which describe the string propagating in a background represented by fields corresponding to the zero mode excitations of the string itself [1] . Typically the pure bosonic part of the sigma-model action will contain the term

$$g^{ab}(h^*\Gamma)_{ab} + \epsilon^{ab}(h^*B)_{ab} \tag{1}$$

integrated over a given Riemann surface S which represents the Euclidean world-sheet of the string: g^{ab} is the inverse of the intrinsic metric over S; $h : S \longrightarrow M$ is an embedding of S into the target space M (the space-time in the string approach), ϵ_{ab} is the two-dimensional antisymmetric tensor, Γ is a metric on the target space and B is a two form on M. It is well known that requiring conformal invariance in the form of vanishing beta functions gives us information about the dynamics of the background fields. Unfortunately information concerning the geometry and the topology of the

---

(*) On leave of absence from Dipartimento di Fisica dell'Università di Padova and Istituto Nazionale di Fisica Nucleare, Sezione di Padova

(**) Work supported in part by: Ministero Pubblica Istruzione (research project on "Geometry and Physics")

target space is hardly obtainable in this way. However pieces of information about the geometry of the target space can be gotten from anomaly cancellation. Of course in order to have anomalies involving background fields, we have to have chiral fermions interacting with background fields. This is what happens for example in the heterotic string sigma model. Again typically we will have in mind a Lagrangian which beside the term (1) contains also:

$$(2) \qquad\qquad <\psi, \partial\!\!\!/_{\bar{h}_0^* A} \psi > + < \lambda, \partial\!\!\!/_{\bar{\bar{h}}_0^* \omega} \lambda >$$

where A and ω are, respectively, connections on a principal gauge bundle with structure group G and on the orthonormal bundle $O_\Gamma M$ over M, while $\partial\!\!\!/_{\bar{h}_0^* A} (\partial\!\!\!/_{\bar{\bar{h}}_0^* \omega})$ denotes the Dirac operator coupled to the pulled-back connections $\bar{h}_0^* A$ $(\bar{\bar{h}}_0^* \omega)$ as explained by the following diagrams [2]

$$(3) \qquad \begin{array}{ccc} h_0^* P & \xrightarrow{\ \bar{h}_0\ } & P \\ \downarrow{\scriptstyle \pi} & & \downarrow{\scriptstyle \pi} \\ S & \xrightarrow{\ h_0\ } & M. \end{array}$$

$$(4) \qquad \begin{array}{ccc} h_0^* O_\Gamma M & \xrightarrow{\ \bar{\bar{h}}_0\ } & O_\Gamma M \\ \downarrow{\scriptstyle \pi} & & \downarrow{\scriptstyle \pi} \\ S & \xrightarrow{\ h_0\ } & M. \end{array}$$

where $\bar{h}_0$ $(\bar{\bar{h}}_0)$ is a bundle map. The field λ is assumed to be a section of the bundle $S^\pm \otimes h^* TM$, where $S^\pm$ is the spinor bundle over S with positive (negative) chirality, and ψ is a section of the vector bundle $S^\mp \otimes h^* V$ with V being a vector bundle associated to P. Notice that ψ and λ have opposite chirality.

The action given by (1) + (2) gives rise to chiral anomalies which, contrary to what occur in field theory [3] , can cancel if certain geometrical restrictions over the target space are fulfilled. This talk is devoted to explaining this qualitatively new phenomenon.

For the sake of simplicity, unless otherwise stated, henceforth we will refer to the diagram (3) , that is a sigma-model in which the world-sheet fermions are coupled to a pulled back gauge connection. In other words, in the fermionic part of the Lagrangian we consider only (3) . The chiral anomaly corresponding to such a model can be constructed in the following way. Referring to the diagram (3) we consider the maps

$$h_0^*P \times \mathcal{G}_{h_0} \xrightarrow{ev_{h_0}} h_0^*P \xrightarrow{\bar{h}_0} P \tag{5}$$

where $\mathcal{G}_{h_0}$ is the group $\mathrm{Aut}_v(h_0^* P)$, that is the group of vertical automorphisms of $h_0^* P$, namely the group of gauge transformations of h_0^*P, and ev_{h_0} is the evaluation map defined by $ev_{h_0}(u, \psi) = \psi(u)$, for any $u \in h_0^* P$ and $\psi \in \mathcal{G}_{h_0}$. [We remark that the construction we are illustrating is valid for a generic sigma-model, in which case one considers a generic map $f_0 : S \longrightarrow M$, not a particular map like an imbedding. We limit ourselves to imbeddings having in mind the string case].

Now in the Lie algebra of G, we consider the polynomial K with two entries which is invariant under the adjoint action of G, and a generic connection ξ on P. We can construct the connection $ev_{h_0}^* \bar{h}_0^* \xi$ on $h_0^* P \times \mathcal{G}_{h_0}$, and, given a fixed connection A_0 on $h_0^* P$, we can consider the 3-form [3]

$$W_K((\bar{h}_0 \circ ev_{h_0})^* \xi, A_0) \tag{6}$$

on $S \times \mathcal{G}_{h_0}$.

[Here $W_K(A, A_0)$ for two generic connections A, A_0 is a basic 3-form such that $dW_K(A, A_0) = K(F, F) - K(F_0, F_0)$, F and F_0 being the curvatures of A and A_0, respectively [3]].

The form (6) contains all the information concerning the properties of an anomaly. If we differentiate it in $S \times \mathcal{G}_{h_0}$ we obtain

$$(d + \delta) W_K((\bar{h}_0 \circ ev_{h_0})^* \xi, A_0) = 0 \tag{7}$$

where d and δ are the exterior differentials in S and $\mathcal{G}_{h_0}$ respectively. The identity (7) splits into several equations according to the order of the form in M and $\mathcal{G}_{h_0}$. In particular at the identity of $\mathcal{G}_{h_0}$, we have

$$\delta\, i_{(\cdot)} W_K(\bar{h}_0^* \xi, A_0) + d\, i_{(\cdot)} i_{(\cdot)} W_K(\bar{h}_0^* \xi, A_0) = 0 \tag{8}$$

which is the consistency condition for the non integrated anomaly $i_{(\cdot)} W_K(\bar{h}_0^* \xi, A_0)$. Here $i_{(\cdot)}$ denotes the map $Z \longrightarrow i_Z$, where Z is any vector field in $\mathrm{Lie}\mathcal{G}_{h_0}$, and i_Z denotes the interior product, which does not act, however, on A_0. The form

$i_{(\cdot)} W_K(\bar{h}_0^* \xi, A_0)$ is the anomaly relevant to the model we are considering. Its coefficient can be calculated through a perturbative expansion or by using the index theorem, but it will not play any rôle in the following.

Contrary to what happens in field theory, in the present context the above anomaly can be cancelled in some cases. A condition for this to occur is $K(F_\xi, F_\xi)$ being in the kernel of the Weil homomorphism, that is

$$K(F_\xi, F_\xi) = dH \tag{9}$$

where H is a 3-form on M, and F_ξ is the curvature of ξ.

One can see immediately that in field theory the condition (9) cannot be satisfied. The reason is that in gauge field theory the diagram analogous to (3) is

$$\begin{array}{ccc} P & \xrightarrow{\hat{f}} & EG \\ \downarrow{\scriptstyle \pi} & & \downarrow{\scriptstyle \pi} \\ M & \xrightarrow{f} & BG. \end{array}$$

where P is the relevant gauge principal bundle BG and EG the classifying space and universal bundle, respectively, f is the classifying map and $\hat{f}$ any bundle map which covers f. In this case the rôle of the connection ξ is played by a universal connection on EG and the analog of eq. (9) is false. In other words the fundamental difference between sigma-models and gauge field theories is that the latter can be regarded as limiting cases of the former when the target space coincides with the classifying space. In turn this property of field theory of being characterized by "universality" properties is a consequence of the requirement of locality [3]. Eq. (9) which would represent a violation of locality in field theory, is allowed in sigma-model and represents simply a (mild) geometrical condition on the target space. We could phrase this by saying that in sigma-models the locality requirement is less restrictive than in field theory.

We are going to show now that, thanks to condition (9) , it is possible to construct a (generalized Wess-Zumino) counterterm which reproduces the anomaly (6) , so that by subtracting it from the quantum action we obtain an anomaly free theory.

To this end let us consider the space of paths over $Imb(S,M)$, the space of imbeddings of S into M, with initial point a fixed imbedding h_0, and denote it by $\mathcal{P}_{h_0} Imb(S,M)$. Next we define the projection $\pi_1 : \mathcal{P}_{h_0} Imb(S,M) \longrightarrow Imb(S,M)$

defined by sending every path into its endpoint. π_1 is a principal fibration with fiber represented by the space $\Omega_{h_0}\, Imb(S,M)$ of loops passing through h_0.

First of all we remark that, by using the parallelism induced by the connection ξ, we can construct a homomorphism [4]

$$\tau : \Omega_{h_0}\, Imb(S,M) \longrightarrow \mathrm{Aut}_v\, h_0^* P. \tag{10}$$

Next consider the diagram

$$\begin{array}{ccccc} h_0^*P \times \mathcal{P}_{h_0}\, Imb(S,M) \approx \pi_1{}^* ev^* P & \xrightarrow{\bar{\pi}_1} & ev^* P & \xrightarrow{\overline{ev}} & P \\ \downarrow & & \downarrow & & \downarrow \\ S \times \mathcal{P}_{h_0}\, Imb(S,M) & \xrightarrow{\pi_1} & S \times Imb(S,M)_{h_0} & \xrightarrow{ev} & M, \end{array} \tag{11}$$

where $Imb(S,M)_{h_0}$ means the component of $Imb(S,M)$ connected to h_0, $\bar{\pi}_1$ and $\overline{ev}$ are the canonical bundle maps corresponding to π_1 and ev, respectively. The isomorphism between $\pi_1{}^* ev^* P$ and

$$h_0^*P \times \mathcal{P}_{h_0}\, Imb(S,M)$$

is induced by the connection ξ, analogously to the homomorphism (10) . Now consider again the connection ξ on P and construct the connection $\bar{\pi}_1 \overline{ev}^* \xi$ on $h_0^*P \times \mathcal{P}_{h_0}\, Imb(S,M)$. Then, if A_0 is a fixed connection on $h_0^* P$, it is easy to show that

$$d(W_K(\bar{\pi}_1 \overline{ev}^* \xi, A_0) - \pi_1^* ev^* H) = 0 \tag{12}$$

due to eq. (9) .

Now we are going to use the contractibility of the space $\mathcal{P}_{h_0}\, Imb(S,M)$ to prove that there exists a two-form β on $S \times \mathcal{P}_{h_0}\, Imb(S,M)$ such that

$$W_K(\bar{\pi}_1 \overline{ev}^* \xi, A_0) - \pi_1^* ev^* H = d\beta. \tag{13}$$

The key point is that any path space is contractible, so that we can construct a retraction

$$r : S \times \mathcal{P}_{h_0}\, Imb(S,M) \longrightarrow S$$

which togheter with the inclusion $i : S \longrightarrow S \times \mathcal{P}_{h_0}\, Imb(S,M)$ forms a map $i \circ r$: $S \times \mathcal{P}_{h_0}\, Imb(S,M) \longrightarrow S \times \mathcal{P}_{h_0}\, Imb(S,M)$, which is homotopic to the identity. So there exists a homotopy operator $\mathcal{H}$ such that

$$1^* - (i \circ r)^* = \mathcal{H} d + d \mathcal{H}. \tag{14}$$

Using eq. (14) one can easily show that

$$\beta = \mathcal{H}(W_K(\bar{\pi}_1 \overline{ev}^* \xi, A_0) - \pi_1^* ev^* H) \tag{15}$$

fulfills eq. (13) .

The generalized Wess-Zumino term is by definition

$$\int_M B^{WZ} \tag{16}$$

where B^{WZ} is the (2,0)-component of β. By construction, if we take the variation of $\int_M B^{WZ}$ along the fiber $\Omega_{h_0}\, Imb(S, M)$, we obtain the (integrated) anomaly of eq. (8) . Of course this identification is permitted if we can identify the variation along the loop space with the variation in the gauge group $\mathcal{G}_{h_0}$. This is accomplished through the homomorphism (10) . Finally, by subtracting the term (16) with a suitable coefficient we can rid the model of the anomaly in question.

Now let us return to the full Lagrangian (1) + (2), which can be suitably described in terms of the bundle $P + O_\Gamma M$ over M. We can consider the diagram (11) with P substituted by $P + O_\Gamma M$. The anomaly is generated by $W_K(\bar{\pi}_1 \overline{ev}^* \xi, A_0) - W_K(\bar{\pi}_1 \overline{ev}^* \eta, \omega_0)$, where A_0, ω_0 are fixed connections on P and $O_\Gamma M$. If eq. (9) and an analogous equation for the connection η is satisfied we can eliminate the two anomalies separately. However there exists also a weaker condition, that is:

$$K(F_\xi, F_\xi) - K(F_\eta, F_\eta) = dH. \tag{17}$$

If eq. (17) is satisfied, in exactly the same way we have proceeded above, we can construct a generalized Wess-Zumino term $\int_M B_{tot}^{WZ}$ which cancels the total anomaly. This is a geometrical description of the Green-Schwarz cancellation mechanism in sigma-models. Needless to say, the term B_{tot}^{WZ} does not have anything to do with the B of eq. (1), except that they are both 2-forms. We can say that B_{tot}^{WZ} represents a very complicated interaction of the bosonic part of the heterotic string with the gauge and Lorentz connections, which annihilate exactly the anomaly generated by the fermionic part. However this is not yet a satisfactory cancellation scheme. Indeed both $\int_M B^{WZ}$ and $\int_M B^{WZ}_{tot}$ are functionals on $\mathcal{P}_{h_0}\, Imb(S,M)$. We have to make sure that the exponential of the product of $2\pi i$ times our (generalized) Wess-Zumino term, descends to a functional over the appropriate degrees of freedom.

A similar situation is met in Witten's bosonized sigma-model [5] where a Wess-Zumino term is added in order to reproduce the features of the original fermionic model. The construction turn out to be a particular case of that described above: the target space M is replaced by a Lie group G, the relevant maps are maps from M to G, $Map(M,G)$ and instead of a generic connection we have the Maurer-Cartan form of G. The construction is simplified and leads to a Wess-Zumino term of the form $\int_{M\times I} Ev^*TK(\theta)$, where Ev is the double evaluation map Ev : $M \times I \times Map_*(M, \mathcal{P}_e G) \longrightarrow G$ and $\mathcal{P}_e G$ is the path space over G wit h initial point the identity. It is easy to realize that

$$exp(2\pi i \int_{M\times I} Ev^*TK(\theta))$$

which is a functional from the space of paths over $Map(M,G)$ to $\frac{\mathbf{R}}{\mathbf{Z}}$, descends to a functional over $Map(M,G)$ only if for any loop $l : M \times S^1 \longrightarrow G$ passing through the identity, the integral $\int_{M\times S^1} l^*TK(\theta)$ is an integer. This requirement gives rise to the coupling quantization condition.

From the previous example it is clear that we have to study for instance the behaviour of the generalized Wess-Zumino term B^{WZ} under gauge transformations which imply changing the homotopy class within $\mathcal{G}_{h_0}$ (whenever $\pi_0(\mathcal{G}_{h_0}) \neq 0$). As a consequence global anomalies become involved. In other words we have to make sure that also global gauge anomalies are eliminated. If this happens, the generalized Wess-Zumino term, when restricted to a functional over the the loop space $\Omega_{h_0}\, Imb(S,M)$, descends to a functional over the subgroup of the group of gauge transformations, given by the image of the homomorphism (10) .

1. References

[1] E.Fradkin and A.Tseytlin:*Effective field theory from quantized string*; Phys.Let. **158B**; 316 (1985);

[2] G.Moore, P.Nelson:*The aetiology of sigma-model anomalies*; Comm.Math.Phys. **100**, 83 (1985);

[3] L.Bonora, P.Cotta-Ramusino, M.Rinaldi and J.Stasheff:*The evaluation map in field theory, sigma-models and strings-I*; CERN-TH 4647/87; to be published in Comm.Math.Phys.;

[4] L.Bonora, P.Cotta-Ramusino, M.Rinaldi and J.Stasheff:*The evaluation map in field theory, sigma-models and strings-II*; in preparation;

[5] E.Witten:*Non abelian bosonization in two dimensions;* Comm.Math.Phys. **92**; 455 (1984).

Lecture Notes in Mathematics

Vol. 1174: Categories in Continuum Physics, Buffalo 1982. Seminar. Edited by F.W. Lawvere and S.H. Schanuel. V, 126 pages. 1986.

Vol. 1184: W. Arendt, A. Grabosch, G. Greiner, U. Groh, H.P. Lotz, U. Moustakas, R. Nagel, F. Neubrander, U. Schlotterbeck, One-parameter Semigroups of Positive Operators. Edited by R. Nagel. X, 460 pages. 1986.

Vol. 1186: Lyapunov Exponents. Proceedings, 1984. Edited by L. Arnold and V. Wihstutz. VI, 374 pages. 1986.

Vol. 1187: Y. Diers, Categories of Boolean Sheaves of Simple Algebras. VI, 168 pages. 1986.

Vol. 1190: Optimization and Related Fields. Proceedings, 1984. Edited by R. Conti, E. De Giorgi and F. Giannessi. VIII, 419 pages. 1986.

Vol. 1191: A.R. Its, V.Yu. Novokshenov, The Isomonodromic Deformation Method in the Theory of Painlevé Equations. IV, 313 pages. 1986.

Vol. 1194: Complex Analysis and Algebraic Geometry. Proceedings, 1985. Edited by H. Grauert. VI, 235 pages. 1986.

Vol. 1203: Stochastic Processes and Their Applications. Proceedings, 1985. Edited by K. Itô and T. Hida. VI, 222 pages. 1986.

Vol. 1209: Differential Geometry, Peñíscola 1985. Proceedings. Edited by A.M. Naveira, A. Ferrández and F. Mascaró. VIII, 306 pages. 1986.

Vol. 1214: Global Analysis – Studies and Applications II. Edited by Yu.G. Borisovich and Yu.E. Gliklikh. V, 275 pages. 1986.

Vol. 1218: Schrödinger Operators, Aarhus 1985. Seminar. Edited by E. Balslev. V, 222 pages. 1986.

Vol. 1227: H. Helson, The Spectral Theorem. VI, 104 pages. 1986.

Vol. 1229: O. Bratteli, Derivations, Dissipations and Group Actions on C*-algebras. IV, 277 pages. 1986.

Vol. 1236: Stochastic Partial Differential Equations and Applications. Proceedings, 1985. Edited by G. Da Prato and L. Tubaro. V, 257 pages. 1987.

Vol. 1237: Rational Approximation and its Applications in Mathematics and Physics. Proceedings, 1985. Edited by J. Gilewicz, M. Pindor and W. Siemaszko. XII, 350 pages. 1987.

Vol. 1250: Stochastic Processes – Mathematics and Physics II. Proceedings 1985. Edited by S. Albeverio, Ph. Blanchard and L. Streit. VI, 359 pages. 1987.

Vol. 1251: Differential Geometric Methods in Mathematical Physics. Proceedings, 1985. Edited by P.L. García and A. Pérez-Rendón. VII, 300 pages. 1987.

Vol. 1255: Differential Geometry and Differential Equations. Proceedings, 1985. Edited by C. Gu, M. Berger and R.L. Bryant. XII, 243 pages. 1987.

Vol. 1256: Pseudo-Differential Operators. Proceedings, 1986. Edited by H.O. Cordes, B. Gramsch and H. Widom. X, 479 pages. 1987.

Vol. 1258: J. Weidmann, Spectral Theory of Ordinary Differential Operators. VI, 303 pages. 1987.

Vol. 1260: N.H. Pavel, Nonlinear Evolution Operators and Semigroups. VI, 285 pages. 1987.

Vol. 1263: V.L. Hansen (Ed.), Differential Geometry. Proceedings, 1985. XI, 288 pages. 1987.

Vol. 1267: J. Lindenstrauss, V.D. Milman (Eds), Geometrical Aspects of Functional Analysis. Seminar. VII, 212 pages. 1987.

Vol. 1269: M. Shiota, Nash Manifolds. VI, 223 pages. 1987.

Vol. 1270: C. Carasso, P.-A. Raviart, D. Serre (Eds), Nonlinear Hyperbolic Problems. Proceedings, 1986. XV, 341 pages. 1987.

Vol. 1272: M.S. Livšic, L.L. Waksman, Commuting Nonselfadjoint Operators in Hilbert Space. III, 115 pages. 1987.

Vol. 1273: G.-M. Greuel, G. Trautmann (Eds), Singularities, Representation of Algebras, and Vector Bundles. Proceedings, 1985. XIV, 383 pages. 1987.

Lecture Notes in Physics

Vol. 282: F. Ehlotzky (Ed.), Fundamentals of Quantum Optics II. Proceedings, 1987. X, 289 pages. 1987.

Vol. 283: M. Yussouff (Ed.), Electronic Band Structure and Its Applications. Proceedings, 1986. VIII, 441 pages. 1987.

Vol. 284: D. Baeriswyl, M. Droz, A. Malaspinas, P. Martinoli (Eds.), Physics in Living Matter. Proceedings, 1986. V, 180 pages. 1987.

Vol. 285: T. Paszkiewicz (Ed.), Physics of Phonons. Proceedings, 1987. X, 486 pages. 1987.

Vol. 286: R. Alicki, K. Lendi, Quantum Dynamical Semigroups and Applications. VIII, 196 pages. 1987.

Vol. 287: W. Hillebrandt, R. Kuhfuß, E. Müller, J.W. Truran (Eds.), Nuclear Astrophysics. Proceedings. IX, 347 pages. 1987.

Vol. 288: J. Arbocz, M. Potier-Ferry, J. Singer, V.Tvergaard, Buckling and Post-Buckling. VII, 246 pages. 1987.

Vol. 289: N. Straumann, Klassische Mechanik. XV, 403 Seiten. 1987.

Vol. 290: K.T. Hecht, The Vector Coherent State Method and Its Application to Problems of Higher Symmetries. V, 154 pages. 1987.

Vol. 291: J.L. Linsky, R.E. Stencel (Eds.), Cool Stars, Stellar Systems, and the Sun. Proceedings, 1987. XIII, 537 pages. 1987.

Vol. 292: E.-H. Schröter, M. Schüssler (Eds.), Solar and Stellar Physics. Proceedings, 1987. V, 231 pages. 1987.

Vol. 293: Th. Dorfmüller, R. Pecora (Eds.), Rotational Dynamics of Small and Macromolecules. Proceedings, 1986. V, 249 pages. 1987.

Vol. 294: D. Berényi, G. Hock (Eds.), High-Energy Ion-Atom Collisions. Proceedings, 1987. VIII, 540 pages. 1988.

Vol. 295: P. Schmüser, Feynman-Graphen und Eichtheorien für Experimentalphysiker. VI, 217 Seiten. 1988.

Vol. 296: M. Month, S. Turner (Eds.), Frontiers of Particle Beams. XII, 700 pages. 1988.

Vol. 297: A. Lawrence (Ed.), Comets to Cosmology. X, 415 pages. 1988.

Vol. 298: M. Georgiev, F' Centers in Alkali Halides. XI, 282 pages. 1988.

Vol. 299: J.D. Buckmaster, T. Takeno (Eds.), Mathematical Modeling in Combustion Science. Proceedings, 1987. VI, 168 pages. 1988.

Vol. 300: B.-G. Englert, Semiclassical Theory of Atoms. VII, 401 pages. 1988.

Vol. 301: G. Ferenczi, F. Beleznay (Eds.), New Developments in Semiconductor Physics. Proceedings, 1987. VI, 302 pages. 1988.

Vol. 302: F. Gieres, Geometry of Supersymmetric Gauge Theories. VIII, 189 pages. 1988.

Vol. 303: P. Breitenlohner, D. Maison, K. Sibold (Eds.), Renormalization of Quantum Field Theories with Non-linear Field Transformations. Proceedings, 1987. VI, 239 pages. 1988.